Routledge Revivals

Metallic Mineral Exploration

How has exploration for minerals evolved in recent years? Is it as productive an activity as it once was? Why have changes occurred? Roderick G. Eggert explores these and other questions about the complex set of circumstances surrounding metallic mineral exploration. Originally published in 1987, Eggert documents trends in the level and the distribution of expenditures by mining companies for metallic mineral exploration and examines a number of factors that may be responsible for these trends. This significant study serves as a handy introduction to the subject for students interested in environmental studies, natural resources, and economics.

Metallic Mineral Exploration

An Economic Analysis

Roderick G. Eggert

First published in 1987
by Resources for the Future, Inc.

This edition first published in 2016 by Routledge
2 Park Square, Milton Park, Abingdon, Oxon, OX14 4RN
and by Routledge
711 Third Avenue, New York, NY 10017

Routledge is an imprint of the Taylor & Francis Group, an informa business

Publisher's Note
The publisher has gone to great lengths to ensure the quality of this reprint but points out that some imperfections in the original copies may be apparent.

Disclaimer
The publisher has made every effort to trace copyright holders and welcomes correspondence from those they have been unable to contact.

A Library of Congress record exists under LC control number: 87042622

ISBN 13: 978-1-138-95077-1 (hbk)
ISBN 13: 978-1-315-66857-4 (ebk)
ISBN 13: 978-1-138-95078-8 (pbk)

METALLIC MINERAL EXPLORATION

AN ECONOMIC ANALYSIS

RODERICK G. EGGERT

A STUDY FROM

Washington, D.C.

Printed in the United States of America

Published by Resources for the Future
1616 P Street, N.W., Washington, D.C. 20036

Books from Resources for the Future are distributed worldwide by The Johns Hopkins University Press

Library of Congress Cataloging-in-Publication Data

Eggert, Roderick G.
Metallic mineral exploration.

Bibliography: p.
1. Prospecting—Costs. I. Title.
TN270.E356 1987 338.2'3 87-42622
ISBN 0-915707-30-6

∞ The paper in this book meets the guidelines for permanence and durability of the Committee on Production Guidelines for Book Longevity of the Council on Library Resources.

Roderick G. Eggert, is assistant professor in the Mineral Economics Department at the Colorado School of Mines.

RESOURCES FOR THE FUTURE (RFF) is an independent nonprofit organization that advances research and public education in the development, conservation, and use of natural resources and in the quality of the environment. Established in 1952 with the cooperation of the Ford Foundation, it is supported by an endowment and by grants from foundations, government agencies, and corporations. Grants are accepted on the condition that RFF is solely responsible for the conduct of its research and the dissemination of its work to the public. The organization does not perform proprietary research.

RFF research is primarily social scientific, especially economic, and is concerned with the relationship of people to the natural environment—the basic resources of land, water, and air; the products and services derived from them; and the effects of production and consumption on environmental quality and human health and well-being. Grouped into three research divisions—Energy and Materials, Quality of the Environment, and Renewable Resources—staff members pursue a wide variety of interests, including food and agricultural policy, water supply and allocation, forest economics, risk management, natural gas policy, multiple use of public lands, mineral economics, air and water pollution, energy and national security, hazardous wastes, climate resources, and the economics of outer space. Resident staff members conduct most of the organization's work; a few others carry out research elsewhere under grants from RFF.

Resources for the Future takes responsibility for the selection of subjects for study and for the appointment of fellows, as well as for their freedom of inquiry. The views of RFF staff members and the interpretations and conclusions of RFF publications should not be attributed to Resources for the Future, its directors, or its officers. As an organization, RFF does not take positions on laws, policies, or events, nor does it lobby.

CONTENTS

PREFACE ix

INTRODUCTION 1

one **THE NATURE OF MINERAL EXPLORATION** 4

Exploration Methods 5
Exploration for Other Natural Resources 13

two **EXPLORATION EXPENDITURES IN SELECTED COUNTRIES** 15

Australia 23
Canada 26
The United States 26
South Africa 29
Developing Countries 30
Summing Up 33
Appendix 2–A. Mineral Exploration in the Soviet Union 34

three **THE EPISODIC NATURE OF EXPLORATION FOR PARTICULAR MINERALS** 36

Iron Ore and Bauxite 37
Copper 40
Molybdenum 43
Uranium 45
Gold 53

four **ECONOMIC PRODUCTIVITY OF MINERAL EXPLORATION** 57

Simple Productivity Measures 60

Success Ratios 62
Discounted Net Returns from Exploration 67

five **CONCLUSIONS 74**

STATISTICAL APPENDIX 77

TABLES

1–1. Principal Discovery Methods for Canadian Metal Deposits, Pre-1920–1975. 6
1–2. Principal Discovery Methods for U.S. Metal Deposits, 1951–70 7
1–3. Activity Breakdown of Mineral Exploration Expenditures by U.S. Companies, 1980–84 9
1–4. Activity Breakdown of Mineral Exploration Expenditures by Canadian Companies, 1980–84 9
1–5. Private Drilling Expenditures in Australia as a Percentage of Total Private Expenditures for Mineral Exploration, 1968/69–1984/85 10
1–6. Private Mineral Exploration Expenditures on Production Leases in Australia as a Percentage of Total Private Expenditures on Mineral Exploration, 1968/69–1984/85 12
1–7. On-Property Expenditures for Mineral Exploration in Canada as a Percentage of Total Mineral Exploration Expenditures, 1968–81 13
1–8. Estimated Exploration Expenditures in the United States, 1979–84 14
1–9. Private Exploration Expenditures in Australia, 1979/80–1984/85 14

2–1. U.S. Share of Total Mineral Exploration Expenditures by Nine Companies, 1971–82 27
2–2. U.S. Share of Total Mineral Exploration Expenditures by U.S. and Canadian Companies, 1980–84 28

3–1. Metallic Mineral Exploration Expenditures in Australia, 1976/77–1984/85 39
3–2. Exploration Drilling in the United States, 1973–82 40
3–3. U.S. Porphyry Copper Deposits Discovered Between 1945 and 1984 41
3–4. U.S. Porphyry Molybdenum Deposits Discovered Between 1940 and 1984 44
3–5. Uranium Exploration: Cumulative Expenditures and Annual Average Intensity in Selected Countries, 1972–83 51
3–6. Two Estimates of Uranium Exploration Expenditures in the United States, 1972–84 52
3–7. Estimates of Reasonably Assured and Estimated Additional Reserves of Uranium, 1967 and 1983 52
3–8. Major Gold Deposits Discovered in the United States, Between 1940 and 1982 56

4–1. Success Ratios for French Exploration, 1973–82 66
4–2. International Comparisons of Success Ratios 68
4–3. Base Metal Exploration Productivity: Base Case Conditions and Calculations for Australia and Canada 71

APPENDIX TABLES

A–1. Historical Mineral Exploration Expenditures, 1960–84 78
A–2. Average Annual Mineral Prices, 1960–85 80
A–3. Uranium Exploration Expenditures by European Companies and in the United States, 1966–84 82
A–4. OECD-IAEA Estimates of Uranium Exploration Expenditures for Selected Countries and the World Outside the Centrally Planned Economies, 1972–83 83
A–5. Gold Exploration Expenditures in Selected Countries and by French Organizations, 1960–83 84

FIGURES

1–1. Typical exploration and development stages 11

2–1. Exploration expenditures for metallic minerals in selected countries and by European companies 16
2–2. Prices for selected metals, 1960–85 19
2–3. Schematic supply and demand schedules for exploration funds 21
2–4. European Community companies: developing country share of exploration expenditures, 1966–84 31

3–1. Discovery of U.S. copper reserves, 1840–1980 42
3–2. European and U.S. uranium exploration expenditures, 1966–84 47
3–3. OECD-IAEA estimates of uranium exploration expenditures, 1972–83 48
3–4. Uranium prices, 1960–85 50
3–5. Gold exploration expenditures in selected countries, 1961–83 54
3–6. Gold's share of exploration expenditures in selected countries and by French explorationists, 1961–83 55
3–7. Gold prices, 1960–85 56

4–1. Canadian mineral deposits discovered per year, 1951–76 61
4–2. Canadian exploration costs per discovery 61
4–3. Canadian discovery values per exploration dollar at January 1979 prices, 1946/48–1981/82 64
4–4. U.S. success ratios, 1955/59–1980/83 65
4–5. Time distribution of average cash flows for an economic deposit in Australia and Canada 70

PREFACE

This study represents part of a larger program of research on the economics of metallic mineral exploration undertaken by the Mineral Economics and Policy Program, which is sponsored by Resources for the Future and the Colorado School of Mines. The research was motivated by the lack of empirical economic analysis in this area and focuses on three sets of questions. First, how have the level and distribution of exploration evolved in recent years? In particular, to what extent has the level of activity declined as a result of the recent financial problems of many of the world's mining companies? Have explorationists avoided searching in many developing countries because of excessive political risks? What changes have occurred in the mix of minerals sought? Second, what factors are responsible for these changes in the level and distribution of exploration? How important, relative to one another, are mineral prices, government policies and political risks, technological advances (in exploration, extraction, and processing), the availability of geologic information, and other factors? Third, how has the effectiveness, or productivity, of exploration changed over time, and why? Has exploration become more expensive because the easier-to-find deposits generally are discovered first?

This study presents an overview and synthesis of the results of the broader program of research. It draws on the other major product of the research program, a series of detailed papers in the volume, *World Mineral Exploration: Trends and Economic Issues* (Washington, D.C., Resources for the Future, 1987), edited by John E. Tilton, Roderick G. Eggert, and Hans H. Landsberg. The present study goes beyond merely summarizing the findings of the edited volume of papers, however, by filling gaps left by these papers and by integrating and analyzing material that until now has been widely scattered. It also will be readable by a wider audience than the edited volume, which in places requires some understanding of ore deposit geology and mining.

This study would not have been possible without the help and support of many people and organizations. In particular, I wish to thank Hans H. Landsberg and John E. Tilton for organizing and leading the larger study of mineral exploration economics, described above, and funded by the Sloan Foundation and the International Institute for Applied Systems Analysis (IIASA), Laxenburg, Austria. For their careful readings of earlier versions of the manuscript, I thank Paul A. Bailly, Phillip Crowson, John H. DeYoung, Jr., Karen Hendrixson, Arthur W. Rose, John J. Schanz, and three anonymous reviewers associated with the RFF Publications Committee. Their comments greatly improved the final product. I, of course, am responsible for any remaining errors or faulty analysis.

Much of the initial research was conducted during a two-year appointment at IIASA. The study was written during one-year appointments at Resources for the Future and at The Pennsylvania State University, both of which provided excellent working environments. Angela Blake and Pat Kustenbauder typed successive versions of the text.

March 6, 1987

Roderick G. Eggert
Colorado School of Mines

INTRODUCTION

During much of the 1970s and early 1980s—in a sort of modified revival of earlier, similar episodes—the specter of mineral shortages and inadequate investment dominated public debate about metallic minerals. In a time-honored tradition, some people argued that the world was in danger of "running out" of minerals as the stock of exploitable mineral deposits was depleted. Others believed that shortages, or at least higher mineral prices, might occur in some ten or twenty years not as a result of a depletion of the world's deposits but rather because of insufficient investment in new mines, processing facilities, and mineral exploration. Onerous government policies and political risks—particularly in developing countries—were alleged to have discouraged exploration in geologically attractive areas.

Today, in the mid-1980s, the fear of mineral shortages has, in the face of low mineral prices, stagnant demand, mine shutdowns, and financial losses of many mining companies, given way to the perception of potentially chronic mineral surpluses and past overinvestment. Some observers suggest that little mineral exploration will be required in the foreseeable future because of the many known but undeveloped deposits. In other words, conventional wisdom about mineral shortages, in general, and the need for mineral exploration, in particular, have changed completely in less than five years.

Through all of this, there has been surprisingly little systematic empirical research documenting trends in the level and distribution of exploration activity and in assessing the factors that have importantly shaped these trends.

The purpose of this study, an historical and economic analysis of metallic mineral exploration, is to partially fill this gap in the literature. The study combines relatively simple economic analysis with technical and institutional aspects of mineral exploration. Although the study makes no attempt to determine if recent and current levels of exploration are too large or too small from society's point of view, it has value for other reasons. It clearly demonstrates that mineral exploration is properly considered an economic activity, which must respond to changes in the underlying factors of mineral supply and demand, rather than an activity whose primary goal is merely to replace reserves at operating mines. Exploration is ultimately motivated by the demand for minerals to be refined and then used in buildings, machinery, consumer durables, and other products. As such it is properly thought of as an investment motivated by the prospect of future economic gain. Changes in the level and distribution of exploration activity, therefore, are importantly shaped by changes in those factors that influence the investment attractiveness of this activity such as mineral prices, government policies, and the quality and availability of geologic information. Furthermore, as will be repeated several times, the study argues that exploration is not the only way that societies respond to the cost-increasing effects of the depletion of known mineral deposits. Material substitution, technological advances in mining and processing, recycling, conservation, and other changes in consumption patterns also counter the effects of depletion.

Although the study draws heavily on more detailed material in Tilton, Eggert, and Landsberg (forthcoming 1987), it goes beyond merely summarizing previous research by synthesizing and analyzing information that until now has been scattered. The coverage is worldwide, but because of data limitations—especially with regard to many developing countries and the centrally planned economies of Eastern Europe, the Soviet Union, and China—it is by no means comprehensive. In some sections the emphasis is decidedly on North American exploration. The study is aimed at a general audience, including mineral economists, exploration geologists, policy analysts, and students.

Chapter 1 discusses the role of exploration in mineral supply, describes various technical aspects of exploration, and examines the evolution of exploration from the prospector and his search for mineralization in surface outcrops to the scientific quest for concealed ore bodies. Chapter 2 traces the level of exploration over time in particular countries and finds that current and recent mineral prices, which transcend national boundaries and reflect the underlying conditions of mineral supply and demand, are largely responsible for similarities among country trends, whereas differences in the country-specific factors, geologic potential and government policies, are largely responsible for differences among countries. Chapter 3 compares and contrasts exploration trends for iron ore, bauxite, copper, molybdenum, uranium, and gold and suggests that exploration and discovery of particular minerals or deposit types is generally episodic over time.

Chapter 4 reviews the most recent studies of mineral exploration effectiveness in Australia, Canada, France, and the United States. Finally, Chapter 5 summarizes important findings of the study and their implications.

one

THE NATURE OF MINERAL EXPLORATION

Mineral exploration is the search for new mineral wealth. In one sense it is a necessary first step in a three-step sequential process of mineral supply. During exploration a number of techniques—including geophysical surveys, geochemical sampling, geologic mapping, and drilling—are used to identify mineral deposits and then evaluate their economic potential. During the second step, development, those mineral deposits that continue to appear economically attractive after exploration are prepared for mining: ore reserves are estimated, a mine and mill are designed and constructed, financing is arranged, infrastructure is provided for, and marketing strategies are prepared. Finally, during the third step of extraction and processing, mineral deposits are mined and the resultant ore, often containing a relatively small percentage of usable metal, is transformed into a purer form that can be fabricated into metal products used by consumers. In this three-step process of mineral supply, one step leads directly to the next. Thus exploration assumes an important role because mineral deposits must be found before they can be developed and mined. Over time known mineral deposits are depleted, and continued exploration is necessary to replace them.

In a broader dynamic sense, however, mineral exploration is only one of several ways that society responds to the cost-increasing effects of the depletion of known mineral deposits, brought about as miners move to lower-grade ore and greater depths below the surface at existing mines and to more remote locations to extract ore at known but previously unexploited deposits. Exploration results in the discovery of lower-cost deposits than otherwise would be available. Other societal responses include material substitution, which acts to replace relatively scarce minerals with relatively abundant ones. During the twentieth century, for example, iron and steel, aluminum, and other materials have substantially replaced wood in construction. Technological advances in mining and mineral processing also place downward pressure on material costs; the flotation process of concentrating sulfide minerals and continuous casting of steel products are only two of many improvements that have lowered unit costs of production. Finally, conservation and recycling, as well as shifts in consumption patterns, have allowed society to stretch out the use of known mineral deposits and thus counter the effects of depletion.

Exploration Methods

Most mineral deposits that have been mined down through history were discovered by direct visual identification of surface mineralization, or, in other words, prospecting. The history and folklore of mining are filled with stories of prospectors roaming the western United States, Canada, and other parts of the world in search of their fortunes. Typically equipped with only an intuitive understanding of geologic science, if that, and little more than grubstake and a pack animal, a prospector relied on his keen eye for surface mineralization.

Organized searches for surface mineral deposits have been undertaken since at least about 4500 B.C. when the Egyptians went to the Sinai Peninsula and the Red Sea to look for copper to be used in jewelry, vases, and weapons (Bailly, 1979). The Romans also are known to have explored for minerals in what is now Spain, Great Britain, and other areas. Most organized searches through the centuries have been conducted close to operating mines and known mineral deposits. In areas with little or no history of mining, chance, or luck, probably has been more important than systematic searches.

In the nineteenth century, core drilling was invented, which permitted subsurface exploration by a method less costly and cumbersome than trenching, tunneling, pitting, or other manual techniques. Core drilling was used to determine, for example, if coal seams in western Europe and Great Britain continued underground from known surface exposures, if lead existed in carbonate rocks in Missouri, and if copper occurred in parts of northern Michigan (Bailly, 1979). Since then exploration has moved away, gradually at first, from direct visual discovery of surface mineral deposits to systematic, scientific identification of deposits concealed below the earth's surface. However, this search for concealed deposits remained the exception rather than the rule until after World War II. By that time core drilling was not the only technique important for discovering concealed mineralization.

Improved geophysical, geochemical, and geologic tools have revolutionized mineral exploration since World War II. Geophysical techniques have been developed to detect magnetic, gravimetric, radiometric, and other physical anomalies associated with certain types of deposits. Geochemical techniques measure chemical anomalies in rocks, soils, stream sediments, and plants. Geophysical and geochemical instrumentation has improved enormously in terms of sensitivity (that is, ability to measure small differences in physical or chemical properties), speed of data collection, sophistication and use of computers in data processing, and miniaturization of equipment. Satellite imagery provides important information on regional geologic structures and rock types. Finally, geologic thinking has focused increasingly on conceptual models of ore occurrence, which relate particular types of mineral deposits to the larger regional geologic environment. These models serve as guides to choosing areas for exploration as well as for subsequent, more detailed evaluations of particular mineral occurrences.

The historical evolution of mineral exploration technology is succinctly summarized in three stages by Harris and Skinner (1982): (1) conventional prospecting of surface geology, (2) detection of shallow deposits using geophysical and geochemical instruments and techniques, and (3) geologic inference using integrated exploration models that in essence predict where concealed ore should occur in areas with little or no surface mineralization or geochemical or geophysical anomalies. Before World War II nearly all mineral discoveries were of the first type. Since then, as more and more of the easy-to-find surface deposits have been detected, discoveries have fit increasingly into the second and third categories, as suggested by data from Canada and the United States (tables 1–1 and 1–2).

Table 1–1. Principal Discovery Methods for Canadian Metal Deposits, Pre-1920–1975

Deposit discovery date	Conventional prospecting		Geophysical anomaly		Geochemical anomaly		Geologic inference		Total number of deposit discoveries
	no.	percent	no.	percent	no.	percent	no.	percent	
Pre-1920	26	93	0	0	0	0	2	7	28
1920–1929	12	80	0	0	0	0	3	20	15
1930–1939	13	87	0	0	0	0	2	13	15
1940–1950	13	76	0	0	0	0	4	24	17
1951–1955	16	46	5	14	0	0	14	40	35
1956–1960	6	25	14	58	0	0	4	17	24
1961–1965	4	27	5	33	2	13	4	27	15
1966–1970	2	10	13	65	1	5	4	20	20
1971–1975	1	4	15	58	3	11	5	19	26[a]

Source: Office of Technology Assessment, Congress of the United States, *Management of Fuel and Nonfuel Minerals in Federal Land: Current Issues and Status* (Washington, D.C., Government Printing Office, 1979), based on D. R. Derry, "Exploration Expenditure, Discovery Rate and Methods," *CIM Bulletin* vol. 63, no. 362 (1970); and D. R. Derry and J. K. B. Booth, "Mineral Discoveries and Exploration Expenditure—A Revised Review 1965–1976," paper prepared for 1977 CIM Symposium.

[a]No principal method listed for two discoveries in 1971–75.

Table 1-2. Principal Discovery Methods for U.S. Metal Deposits, 1951-70

Deposit discovery date	Conventional prospecting		Geophysical anomaly		Geochemical anomaly		Geologic inference		Total number of deposit discoveries
	no.	percent	no.	percent	no.	percent	no.	percent	
1951-1955	1	8	2	17	0	0	9	75	12
1956-1960	2	13	2	13	1	7	10	67	15
1961-1965	0	0	2	13	0	0	13	87	15
1966-1970	0	0	2	11	2	10	15	79	19

Source: Office of Technology Assessment, Congress of the United States, *Management of Fuel and Nonfuel Minerals in Federal Land: Current Issues and Status* (Washington, D.C., Government Printing Office, 1979), based on Paul Bailly, "Changing Rates of Success in Metallic Exploration," paper presented at the Geological Association of Canada-Mining Association of Canada-Society of Exporation Geologists-Canadian Geophysical Union Annual Meeting, Vancouver, British Columbia, April 25, 1977.

There are, however, two important caveats to this idealized sequence. First, not all countries or geographic locations are at the same point in the evolution. Some areas, such as the western United States and much of Europe, have been searched relatively intensely, reducing the likelihood that conventional prospecting will play much of a role in future discoveries. But in more remote areas, where relatively little prospecting has occurred, direct visual identification of surface mineralization still may be important. Second, not all minerals or deposits are at the same stage. Most iron ore and bauxite discoveries since World War II, for example, fit into the first two stages, and it is unlikely that iron ore and bauxite will move to the third stage any time soon, if ever. Many known mineral occurrences await detailed evaluation, and over the long term society may extract ore from progressively lower-grade mineralization rather than call on sophisticated geologic inference to locate new deposits. On the other hand, most base metals exploration, especially for volcanogenic massive sulfide, porphyry copper, and molybdenum deposits, has been in the second and third stages for a decade or more.

What follows is a generalized sequence of activities typical of modern mineral exploration. Although directly applicable to private organizations, the sequence and its activities are applicable with only slight modifications to public enterprises. It begins with the *idea stage,* in which two fundamental questions are answered: what to explore for and where to explore. In other words, explorationists choose minerals, deposit types, and geographic locations for exploration. Typical activities include a library search of the geographic literature, economic and political analysis, corporate planning to identify corporate goals, and perhaps initial fieldwork.

During the second stage, *reconnaissance,* explorationists examine large areas of land to select small targets for subsequent examination. This is a sifting process, the goal of which is to narrow the focus of the search and to improve the odds of discovering mineralization that can be mined for a profit. Large areas with low potential for ore mineralization are identified and eliminated from further consideration. Typical exploration activities include: airborne surveys to identify geophysical anomalies; regional sampling of rocks, soil, water, and plant life to detect geochemical anomalies; large-scale geologic mapping to identify surface mineralization or associated alteration zones as well as to follow up any regional geophysical or geochemical anomalies; and acquisition of the mineral rights to promising targets. The particular methods used by an explorationist will vary according to the specific deposit type and physical characteristics of the exploration area.

During the final stage, *target evaluation and discovery*, the focus of the search narrows further. The goal is to identify mineral deposits in the targets chosen during the previous stage and then to evaluate the deposits in sufficient detail to make a "go–no go" decision on mine development. Typical activities begin with surface tests, including ground geophysical surveys, detailed geochemical sampling, and detailed geologic mapping. Exploration then shifts to direct examination of subsurface rocks, primarily through drilling, based on geologic hypotheses developed from the surface data. If the results of initial drilling holes are favorable, a program of detailed drilling may be undertaken to delineate a mineral deposit and to evaluate its economic potential. Drilling costs constitute a sizable portion of overall exploration costs—for example, between about one-fifth and two-fifths of total costs in the United States, Canada, and Australia for recent years (tables 1–3, 1–4, and 1–5). Economic evaluation becomes increasingly important at this stage of exploration. A feasibility study assesses all relevant economic and technical factors determining the commercial viability of a deposit and usually serves as the basis for deciding whether or not to proceed with mine development.

Several comments need to be made about this idealized sequence of exploration activities. First, each successive stage typically involves more time, more money, and less land than the previous stage (figure 1–1). Moreover, discovery risk declines; that is, the odds increase for discovering a commercial mineral deposit as exploration moves from initial fieldwork and reconnaissance to target evaluation. Second, mineral exploration is essentially a means of collecting information. At each stage an explorationist collects technical and economic information to narrow the focus of his search and to increase the probability of making an economic discovery. At the idea stage, an explorationist gathers information to select minerals, deposit types, and regions for exploration; at the target selection (reconnaissance) stage to eliminate broad areas with low potential for commercial mineralization and to select targets for subsequent detailed examination; and at the target

Table 1–3. Activity Breakdown of Mineral Exploration Expenditures by U.S. Companies, 1980–84[a]

Activity	1980 1,000 US$	1980 percent	1982 1,000 US$	1982 percent	1984 1,000 US$	1984 percent
Drilling	164,732	32.1	145,835	28.0	83,687	27.1
Airborne geophysics	8,658	1.7	8,309	1.6	5,444	1.8
Ground geophysics	21,889	4.3	22,310	4.3	8,162	2.6
Geochemistry & prospecting	105,248	20.5	104,589	20.1	59,501	19.2
Land acquisition	56,925	11.1	44,157	8.5	21,794	7.1
Other	66,917	13.0	96,659	18.6	69,173	22.4
General administration	88,694	17.3	98,314	18.9	61,372	19.9
Total[b]	513,063	100.0	520,173	100.0	309,133	100.1[c]

Note: Dollar figures based on current prices.

Source: Calculated from data in G. A. Barber and S. Muessig, "Minerals Exploration Statistics for the Years 1980, 1981, and 1982," *Economic Geology* vol. 79, no. 7 (1984) pp. 1768–1776; and G. A. Barber and S. Muessig, "1984 Minerals Exploration Statistics: United States and Canadian Companies," *Economic Geology*, in press.

[a]Based on fifty-nine reporting companies in 1980 and 1982 and forty-eight reporting companies in 1984.

[b]Includes expenditures for base and precious metals, other metals, uranium, and industrial minerals. Excludes expenditures for oil, gas, and coal. Excludes drilling expenditures aimed at expanding ore reserves delineated already at existing mines, development drilling, pre-engineering drilling, and drilling to acquire metallurgical samples.

[c]Totals may not sum to 100 percent because of rounding.

Table 1–4. Activity Breakdown of Mineral Exploration Expenditures by Canadian Companies, 1980–84[a]

Activity	1980 1,000 US$	1980 percent	1982 1,000 US$	1982 percent	1984 1,000 US$	1984 percent
Drilling	69,591	36.4	49,261	30.2	62,594	39.1
Airborne geophysics	5,247	2.7	3,544	2.2	1,886	1.2
Ground geophysics	24,981	13.1	15,071	9.2	13,740	8.6
Geochemistry & prospecting	26,979	14.1	20,563	12.6	14,356	9.0
Land acquisition	9,681	5.1	7,365	4.5	4,653	2.9
Other	31,078	16.3	41,307	25.3	39,387	24.6
General administration	23,450	12.3	26,026	16.0	23,645	14.8
Total[b]	191,007	100.0	163,137	100.0	160,261	100.2[c]

Note: Dollar figures based on current prices.

Source: Calculated from data in G. A. Barber and S. Muessig, "Minerals Exploration Statistics for the Years 1980, 1981, and 1982," *Economic Geology* vol. 79, no. 7, (1984) pp. 1768–1776; and G. A. Barber and S. Muessig, "1984 Minerals Exploration Statistics: United States and Canadian Companies," *Economic Geology*, in press.

[a]Based on 59 reporting companies in 1980 and 1982 and 48 reporting companies in 1984.

[b]Includes expenditures for base and precious metals, other metals, uranium, and industrial minerals. Excludes expenditures for oil, gas, and coal. Excludes drilling expenditures aimed at expanding ore reserves delineated already at existing mines, development drilling, pre-engineering drilling, and drilling to acquire metallurgical samples.

[c]Totals may not sum to 100 percent because of rounding.

Table 1–5. **Private Drilling Expenditures in Australia as a Percentage of Total Private Expenditures for Mineral Exploration, 1968/69–1984/85**

Fiscal year (July 1–June 30)	Drilling expenditures (million A$)	Total expenditures[a] (million A$)	Share (percent)
1968/69	24.37	72.56	33.59
1969/70	32.06	118.11	27.14
1970/71	43.15	161.06	26.79
1971/72	30.94	117.06	26.43
1972/73	26.80	99.73	26.87
1973/74	25.93	101.63	25.51
1974/75	32.88	109.83	29.94
1975/76	30.53	99.98	30.54
1976/77	36.14	133.97	26.98
1977/78	50.02	158.39	31.58
1978/79	50.73	182.51	27.80
1979/80	72.41	286.13	25.31
1980/81	126.09	470.49	26.80
1981/82	141.87	575.57	24.65
1982/83	89.72	437.91	20.49
1983/84	93.50	428.70	21.81
1984/85	96.73	437.33	22.12

Note: Dollar figures based on current prices.

Sources: 1968/69–1981/82: Based on P. C. F. Crowson, "A Perspective on Worldwide Exploration for Minerals," in John E. Tilton, Roderick G. Eggert, and Hans H. Landsberg, eds., *World Mineral Exploration: Trends and Economic Issues* (Washington, D.C., Resources for the Future, forthcoming 1987); 1982/83–1984/85: Australian Bureau of Statistics, *Mineral Exploration Australia* (1984–1986), Catalogue no. 8407.0.

[a]The total includes expenditures on production leases of operating mines and development properties, as well as exploration elsewhere.

evaluation and discovery stage to select drilling sites and to assess the commercial viability of the target.

Third, this is a full sequence of activities, and an organization can bypass one or all of the stages if it acquires a promising target for detailed exploration, a known deposit for development, developed reserves for mining, or an operating mine. Some mining companies eschew early stage exploration, choosing instead to rely on acquiring known deposits or developed reserves. Other companies prefer to generate their own mineral prospects and thus are active in the full sequence of activities. It is a question of company strategy.

Fourth, an area can never be explored in a once-and-for-all manner, despite what may appear to be considerable scientific rigor in the three-stage sequence. Reappraisal of previously explored land has led to many significant discoveries because changes in technology have improved the equipment available to explorationists; changes in the geologic sciences have led to revised search strategies; and changes in mineral prices and capital and production costs—often the result of cost-reducing technological improvements in mining and mineral

Time (years)	Stage	Size of area (maximum in 1,000 km²)	Typical range of expenditures (million US$)
−2	1. Idea	1,000	up to 0.3
	• literature search		
−1	• selection of favorable areas		
	• initial field work		
0	2. Reconnaisance, regional geologic, geochemical, and geophysical surveys	100	0.4–1.2
1			
2			
	3. Target evaluation and discovery		
3	• surface examinations	10	2.5–50
5	• 3D subsurface work	1	2.5–50
6	• feasibility study	<1	5.0–25
7	4. Mine		
9	development and construction	<1	
11			
	5. Production: 10 to 30 years		

Figure 1–1. Typical exploration and development stages. Although originally based on uranium exploration, the figure is generally representative of most metallic mineral exploration. The numbers on the left (years) refer to the adjacent time bars. *Source*: Adapted from P.C.F. Crowson, "A Perspective on Worldwide Exploration for Minerals," in John E. Tilton, Roderick G. Eggert, and Hans H. Landsberg, eds., *World Mineral Exploration: Trends and Economic Issues* (Washington, D.C., Resources for the Future, forthcoming 1987).

processing—have altered the definition of what is commercial mineralization and, in effect, converted worthless rock into ore or vice versa. Moreover, and perhaps most importantly, successful exploration is just as much an art as it is a science. It is somewhat akin to solving a murder mystery. The search is based on bits and pieces of seemingly unrelated and incomplete evidence. Different explorationists draw different inferences from the same piece of information.

There are false leads and blind alleys. Many explorationists give up the chase prematurely, while others doggedly pursue something that just does not exist. Luck also may play a role.

Fifth, the three-stage scheme omits two important components of exploration and discovery, as noted by Crowson (forthcoming 1987). It fails to recognize the important role of geologic surveys, usually conducted by government organizations, in providing basic geologic maps and other background information on which formal exploration is based. And it does not include exploration at or near operating mines, which is a very important source of new mineral reserves, even though it often is excluded from exploration statistics. Two countries, however, provide statistics on this type of exploration. In Australia exploration on production leases has accounted for between 9 and 18 percent of annual total exploration expenditures by private organizations since 1968 (table 1–6). In Canada exploration for new

Table 1–6. Private Mineral Exploration Expenditures on Production Leases in Australia as a Percentage of Total Private Expenditures on Mineral Exploration, 1968/69–1984/85

Fiscal year (July 1–June 30)	Production lease expenditures[a] (million A$)	Total expenditures[b] (million A$)	Share (percent)
1968/69	10.18	72.56	14.03
1969/70	21.09	118.11	17.86
1970/71	27.55	161.06	17.11
1971/72	21.18	117.06	18.09
1972/73	13.70	99.73	13.74
1973/74	14.36	101.63	14.13
1974/75	16.76	109.83	15.26
1975/76	15.13	99.98	15.13
1976/77	23.93	133.97	17.86
1977/78	22.82	158.39	14.41
1978/79	23.96	182.51	13.13
1979/80	30.53	286.13	10.67
1980/81	45.25	470.49	9.62
1981/82	44.59	575.57	7.75
1982/83	49.58	437.91	11.32
1983/84	40.67	428.70	9.49
1984/85	44.67	437.33	10.21

Note: Australian dollar figures based on current prices.

Sources: 1968/69–1981/82: Based on P. C. F. Crowson, "A Perspective on Worldwide Exploration for Minerals," in John E. Tilton, Roderick G. Eggert, and Hans H. Landsberg, eds., *World Mineral Exploration: Trends and Economic Issues* (Washington, D.C., Resources for the Future, forthcoming 1987) table Part 2-A-2. Based on data from the Australian Bureau of Statistics; 1982/83–1984/85: Australian Bureau of Statistics, *Mineral Exploration Australia (1984–1986)*, Catalogue no. 8407.0.

[a]Exploration by private organizations on production leases of operating mines and development properties.

[b]Includes coal and nonmetallic minerals but excludes oil and gas.

Table 1–7. On-Property Expenditures for Mineral Exploration in Canada as a Percentage of Total Mineral Exploration Expenditures, 1968–81

Year	On-property expenditures[a] (million C$)	Total expenditures[b] (million C$)	Share (percent)
1968	31.6	105.5	30.0
1969	37.9	136.7	27.7
1970	25.7	144.5	17.8
1971	27.2	118.5	23.0
1972	16.6	89.1	18.6
1973	23.2	110.3	21.0
1974	25.2	136.0	18.5
1975	25.7	154.8	16.6
1976	35.4	166.5	21.3
1977	73.8	243.8	30.3
1978	54.1	249.0	21.7
1979	49.2	298.1	16.5
1980	85.4	489.0	17.5
1981	134.5	626.3	21.5

Note: Canadian dollar figures in current prices.

Source: Based on data from Canada's Energy, Mines, and Resources Department in P. C. F. Crowson, "A Perspective on Worldwide Exploration for Minerals," in John E. Tilton, Roderick G. Eggert, and Hans H. Landsberg, eds., *World Mineral Exploration: Trends and Economic Issues* (Washington, D.C., Resources for the Future, forthcoming 1987) table Part 2-A-7.

[a]Exploration for new ore bodies on properties either in production or being prepared for production.

[b]Includes metallic and nonmetallic minerals. Includes coal but excludes oil and gas.

ore bodies on properties in production or being prepared for production represented between 16 and 30 percent of annual total expenditures between 1968 and 1981 (table 1–7). Comparable data for the United States are not available.

Exploration for Other Natural Resources

Before turning to the analytical sections of the study, it is worthwhile to compare the magnitude of metals exploration spending with outlays for other natural resources. Data are not available for the world as a whole, but they are available for the United States and Australia (tables 1–8 and 1–9). In the United States, oil and gas exploration have persistently exceeded total exploration for all other minerals, and metals exploration, in turn, is far greater than exploration for industrial minerals. In Australia the pattern is broadly similar, although oil and gas exploration dwarf other exploration expenditures by much less than in the United States; also, coal and diamond exploration are broken out in the Australian numbers.

Table 1–8. Estimated Exploration Expenditures in the United States, 1979–84
(million US$, current prices)

	1979	1980	1981	1982	1983	1984
Metals, excluding uranium[a]	190	250	365	305	305	220
Uranium[b]	316	267	145	74	37	26
Industrial minerals[c]	n.a.	23	33	36	18	10
Oil and gas[d]	7,750	10,725	17,125	16,775	11,975	12,500

Note: n.a. = not available.

[a]1980–1983: Estimated with data from G. A. Barber and S. Muessig, "Minerals Exploration Statistics for the Years 1980, 1981, and 1982," *Economic Geology* vol. 79, no. 7 (1984) pp. 1768–1776; Barber and Muessig, "1983 Minerals Exploration Statistics: United States and Canadian Companies," *Economic Geology* vol. 80, no. 7 (1985) pp. 2060–2066; and Barber and Muessig, "1984 Minerals Exploration Statistics: United States and Canadian Companies," *Economic Geology*, in press. Adjusted upward to account for nonreporting companies; 1979: author estimates.

[b]Energy Information Administration, U.S. Department of Energy, *Uranium Industry Annual 1984*, DOE/EIA-0478(84) (Washington, D.C., Government Printing Office, 1985).

[c]Estimated with data from G. A. Barber and S. Muessig, "Minerals Exploration Statistics for the Years 1980, 1981, and 1982," *Economic Geology* vol. 79, no. 7 (1984) pp. 1768–1776; Barber and Muessig, "1983 Minerals Exploration Statistics: United States and Canadian Companies," *Economic Geology* vol. 80, no. 7 (1985) pp. 2060–2066; and Barber and Muessig, "1984 Minerals Exploration Statistics: United States and Canadian Companies," *Economic Geology*, in press. Adjusted upward to account for nonreporting companies.

[d]David J. Behling, Jr., Richard S. Dobias, and Norma J. Anderson, 1984 *Capital Investments in the World Petroleum Industry* (New York, Chase Manhattan Bank). Includes capitalized dry hole expenditures, geological and geophysical expenses, and lease rental expenses. Excludes capitalized lease acquisition expenditures.

Table 1–9. Private Exploration Expenditures in Australia, 1979/80–1984/85
(million A$, current prices)

	1979/80	1980/81	1981/82	1982/83	1983/84	1984/85
Metals, excluding uranium	162.57	291.50	340.67	281.13	329.86	352.29
Uranium	38.26	48.50	56.37	36.55	20.43	13.11
Coal	46.71	74.94	108.73	61.42	43.67	34.57
Diamonds	32.59	51.39	63.11	51.20	26.90	28.58
Other nonmetallics	6.00	7.84	6.69	7.62	7.85	8.78
Oil and gas	294.71	368.33	803.98	927.41	823.69	n.a

Note: n.a. = not available. Years are fiscal years.

Source: P. C. F. Crowson, "A Perspective on Worldwide Exploration for Minerals," in John E. Tilton, Roderick G. Eggert, and Hans H. Landsberg, eds., *World Mineral Exploration: Trends and Economic Issues* (Washington, D.C., Resources for the Future, forthcoming 1987); and Australian Bureau of Statistics, *Mineral Exploration Australia* (1984–1986), Catalogue no. 8407.0.

two

EXPLORATION EXPENDITURES IN SELECTED COUNTRIES

This study turns now to an historical analysis of the level and distribution of worldwide mineral exploration during the last twenty years or so. One way of organizing this analysis would be to look first at changes in the overall level of worldwide exploration—the size of the exploration pie, if you will—and then examine the distribution according to both geographic location and mineral sought—that is, how the pie is sliced. Not surprisingly the available data are insufficient for this task. Instead the analysis is somewhat more limited and is organized as follows: Chapter 2 examines changes in the level of exploration activity in selected countries for which historical data are available and assesses the relative importance of the various factors determining these changes to explain the similarities and differences among countries. Chapter 3 looks at how exploration has varied over time for particular minerals, such as iron ore, copper, and gold, and suggests that exploration for particular minerals is generally episodic.

Figure 2–1 shows historical exploration expenditures (inflation adjusted) in selected countries as well as worldwide expenditures by a group of European companies. The data, although broadly similar, are not completely comparable. The Australian and Canadian data are collected annually by government departments in these countries; the Australian data include all expenditures, including overhead, whereas the Canadian numbers do not include overhead and other indirect costs. The South African and United States numbers are from recent surveys conducted by private researchers. The data for European companies are not country data at all but rather are company data that have been collected annually since the mid-1970s, first by a group of European mining companies and since 1979 by the Comité de Liaison des Industries de Métaux Non Ferreux. In 1976 the number of companies included in the survey increased. For the most part, the

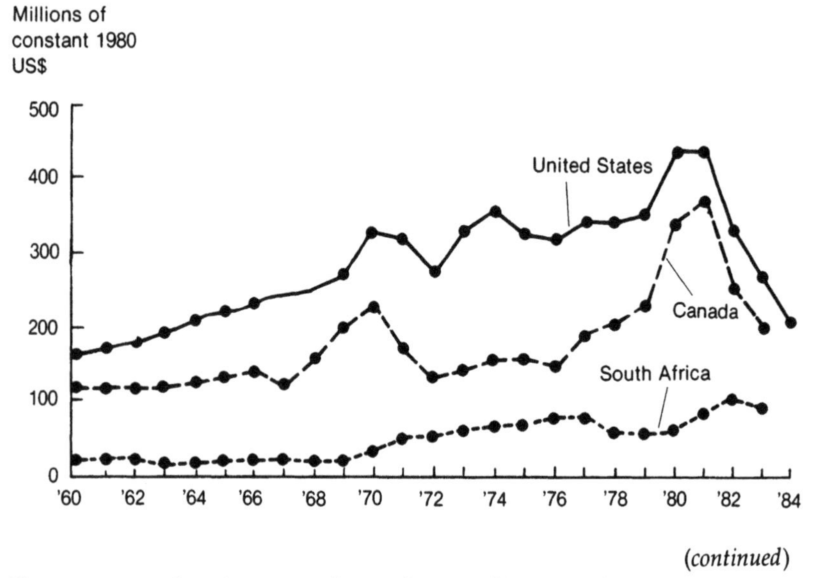

(*continued*)

Figure 2–1. Exploration expenditures for metallic minerals in selected countries and by European companies. For Australia the 1965–67 data are for calendar years; the data for 1968/69–1984/85 are for fiscal years. For the European companies, the expenditure trend is discontinuous because the composition of the companies included in the survey changed in 1976; expenditures for both company bases are included in the figure for 1976. *Source:* appendix table A–1 and author estimates.

Figure 2–1. *(continued)*

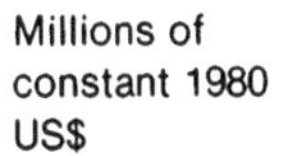

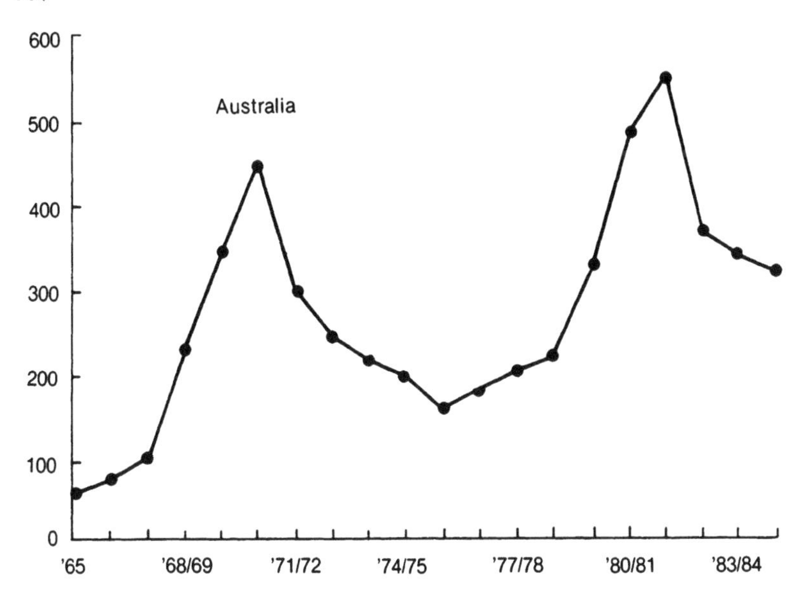

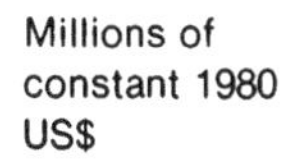

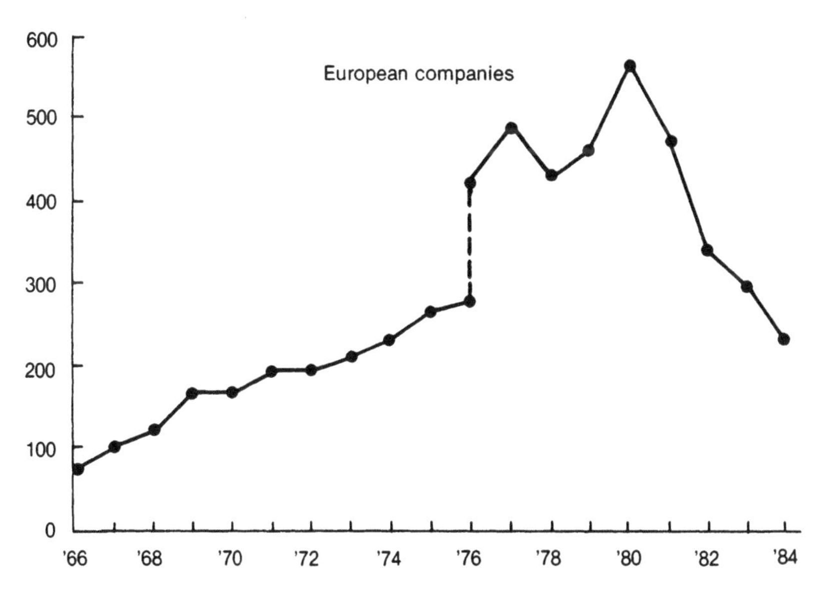

data in figure 2–1 represent expenditures for *metallic* mineral exploration (including uranium), but differences do exist among countries.

Several broad similarities are readily apparent among the expenditure trends. First, the long-term trend in real (inflation-adjusted) expenditures has been upward, but there has been a sharp fall during the 1980s. Second, exploration expenditures have been cyclical. This is particularly true for Australia and Canada but less so for the U.S. and European companies. Expenditure peaks generally occurred in 1970–71, the mid-1970s, and the early 1980s, while expenditures tended to fall in the intervening years.

Despite these similarities—upward trending and cyclical outlays—obvious differences among countries exist. Australia, for example, experienced an extended downturn in expenditures in the mid-1970s while exploration elsewhere was increasing. In South Africa no expenditure peak occurred in the early 1970s, and expenditures increased much more between the late 1960s and early 1980s than expenditures elsewhere. Exploration by the European companies and by thirty companies in the United States shows relatively little year-to-year variation. What accounts for the similarities and differences among historical trends in metallic mineral exploration?

The history of recent mineral prices, which are generally set in world markets and thus transcend national and corporate boundaries, may be responsible for many of the similarities for two reasons. First, changing prices alter expectations about future prices and in turn alter expected revenues and net financial returns from exploration and future mining; the rationale is that expectations about the future are strongly influenced by the present and recent past. Second, mineral prices are an important determinant of mining company profitability and to the extent that exploration is funded out of internal funds rather than through debt or equity they may influence the availability and cost of capital. In both cases rising (falling) prices will lead to increased (decreased) exploration expenditures. Expenditures are likely to follow price changes with a short lag—six months to a year—because next year's expenditure is largely a function of a budgeting process influenced by this year's expectation about future mineral prices and level of internal funds.

Figure 2–2 shows historical prices (inflation-adjusted) for several important metals as well as a price index for primary nonferrous metals. A comparison of mineral prices (figure 2–2) with exploration expenditures (figure 2–1) shows that many fluctuations in expenditures correspond to price variations. Prices increased steadily during the late 1960s, peaking in 1970; during this period exploration rose in Australia, Canada, and the United States. Exploration by the European companies also increased. Expenditures peaked in Canada

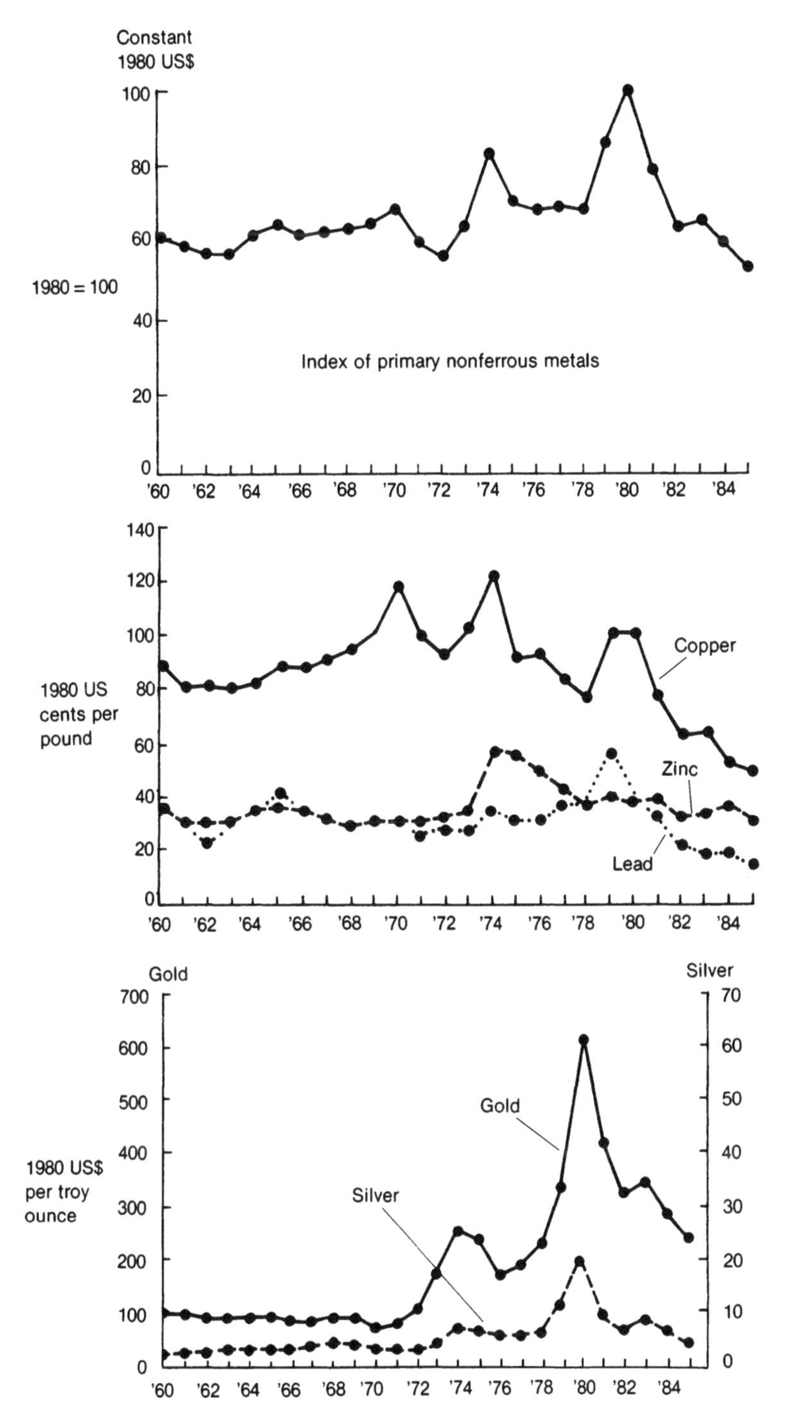

Figure 2–2. Prices for selected metals, 1960–85. *Source:* appendix table A–2.

and the United States in 1970, and in Australia in fiscal year 1970–71. Mineral prices fell during 1971 and 1972, as did exploration expenditures in Australia, Canada, and the United States. The substantial price increases of 1973 and 1974 were followed by increased exploration in Canada, South Africa, and the United States during the mid-1970s. The next price surge occurred from 1979 to 1981, and exploration expenditures hit all-time highs around the world in the early 1980s. By 1985 prices had fallen precipitously, as had exploration expenditures.

As briefly suggested earlier, recent mineral prices may be an important determinant of exploration expenditures for two very different reasons. First, mineral prices reflect the interaction of the underlying supply and demand for minerals and thus strongly influence expected revenues from exploration. Current prices importantly shape expectations of future prices and revenues from mining and in turn shape expected revenues and net returns from exploration. Although a number of factors influence estimates of potential returns from future exploration and mining, including expected development and mining costs and mineral demand, we only argue here that recent prices are a very important determinant of such estimates. Many of the available price-forecasting techniques depend on recent price trends; other techniques may implicitly assume a correlation or other type of relationship between recent and future price trends, and they generally assume that the underlying behavioral relationships governing price determination are maintained over time.

The second reason that exploration expenditures may be related to recent mineral prices is that mineral prices strongly influence the availability and cost of capital for financing exploration by a private company. Mineral prices are one important determinant of mining revenues and, in turn, the availability of internal funds for exploration. Rising prices lead to increased profits and larger pools of internal funds; falling prices have the opposite effect. Prices influence the cost of capital if the costs of external funds (debt or equity) are greater than the cost of internal funds. Figure 2–3 shows how this might work.

SS_1 is a supply curve for investment funds, and the cost of using the limited amount of internal funds is approximately constant. As a firm calls on external sources of debt and equity for additional financing, however, cost of capital increases as the firm's debt/equity ratio increases, or as additional stocks dilute management's control over the firm (see Branson, 1979). *DD* is a demand curve for exploration funds. During periods of low prices and corresponding low levels of internal funds, the equilibrium exploration expenditure corresponds to q_1. During periods of higher prices and correspondingly higher levels of internal funds, the supply schedule shifts to the right,

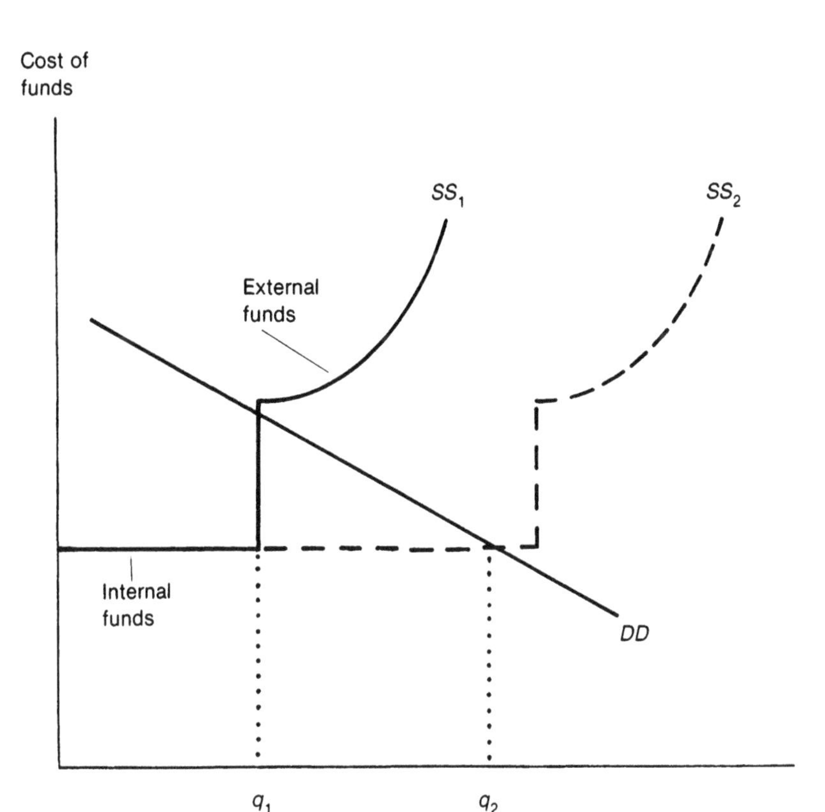

Figure 2–3. Schematic supply and demand schedules for exploration funds

as depicted by SS_2. In this case expendiures increase to q_2, even though the demand curve for exploration funds remains stationary. (Note that in the first argument concerning the relationship between mineral prices and exploration expenditures, mineral price changes cause a shift in the demand curve for exploration funds by altering expected exploration revenues.) Furthermore, corporate exploration involves a great deal of rivalry among competing exploration groups, and companies may be reluctant to call on external sources of finance if it means divulging information about the exploration program that they would prefer to keep confidential.

The preceding theoretical argument provides a rationale for the conventional wisdom that internal funds are the primary source of funds for exploration by large North American firms. According to Paul Bailly (former president of Occidental Minerals, personal communication), the availability of internal funds basically controls exploration by large U.S. companies which do not use any external funds for exploration; only a few small stock companies finance exploration with money raised in regional markets (such as the Vancouver, Toronto, Denver, and Spokane stock exchanges).

Therefore mineral prices appear to influence exploration expenditures in two very different ways—by influencing potential revenues and thus the demand for exploration funds and by altering costs (through their effect on levels of internal funds) and thus the supply of exploration funds. In both cases rising (falling) prices stimulate (discourage) exploration.

If year-to-year changes in mineral prices account for many of the similarities among historical trends in exploration expenditures in particular countries, what factors are responsible for the differences in trends? To answer this question one needs to look at factors that vary from country to country. One obvious factor is geologic potential, or the relative likelihood that a country contains undiscovered or unevaluated mineral deposits that can be mined profitably. Countries or regions within a country with a history of mining usually are perceived to have a higher geologic potential than areas with little or no previous mining. But perceptions of geologic potential for discovery change over time as the understanding or science of ore deposit occurrence evolves and as discoveries are made in areas previously thought to have little potential for ore occurrence. These perceptions are also influenced by changes in relative mineral prices and the costs of mining and mineral processing for particular minerals and deposit types. (For example, some countries have rocks that are particularly attractive for gold mineralization and therefore may experience increased exploration when gold prices rise, even if other mineral prices—and perceptions of future prices—generally fall.)

Although favorable geology is obviously a prerequisite for exploration in a country, it is frequently argued that changing government policies and political risks are largely responsible for the changing geographic allocation of exploration expenditures. During much of the 1970s and early 1980s, for example, it was argued that mining companies virtually ceased exploring in many geologically attractive developing countries because of excessive political risks. At the same time in the developed economies of Australia, Canada, and the United States, mineral exploration was alleged to have been severely hampered by unduly onerous government policies, particularly tax and land regulations.

The rationale is that changes in government policy raise or lower the costs and thus the investment attractiveness of operating in a country by providing incentives or disincentives for exploration, mining, and mineral processing. Incentives take a number of forms, including tax holidays for new mines, exclusive exploration rights for large blocks of land, and concessionary power rates for mineral-processing facilities. Disincentives include host-county ownership requirements, restrictive land and tax policies, and environmental regulations.

Political risk, on the other hand, reflects the possibility that future financial returns will be adversely affected by government actions. Changes in perceived political risk alter the investment attractiveness of a country by raising or lowering the risk premium (over and above a riskless rate of return) required by firms to invest in an area. Although changes in government policies and perceived political risks are conceptually quite different, in practice it is nearly impossible to distinguish between the effects of these two factors. Policy changes that alter exploration and mining costs often simultaneously influence risk perceptions by implying that current policies are not stable and thus cannot be relied on for planning purposes.

The rest of this chapter examines the extent to which government policies and political risks, as opposed to geologic potential, account for the differences among country exploration trends. The Australia and South Africa sections provide broad, historical overviews of exploration in those countries, whereas the Canadian, American, and developing country sections look narrowly at the influence of government policies on exploration.

Australia

Mineral exploration in Australia, as noted earlier, experienced two pronounced peaks and one extended trough between the late 1960s and the early 1980s. It is often alleged that onerous government policies were largely responsible for the length and severity of the exploration downturn in the mid-1970s. These policies, usually associated with the Labour Party, which came to power in 1972 and was defeated in 1975, are supposed to have discouraged exploration in two ways. First, the direct effects of new policies made exploration and mining more costly, time consuming, and difficult by, for example, placing greater restrictions on foreign-owned equity, establishing the Takeover Review Board to approve arrangements between Australian resource owners and foreigners, requiring the consent of aboriginal people to work on their land in the future, and effectively banning further development of most uranium deposits. Second, the indirect effect of increased uncertainty about future policy changes (that is, increased political risk) added to the deterrents.

At first glance the argument that policies significantly discouraged exploration in Australia appears plausible. Constant-dollar expenditures in Australia fell by more than 60 percent between fiscal years

1970–71 and 1975–76 (July 1–June 30) compared with peak-to-trough declines of about 40 percent in Canada and less than 20 percent in the United States. Australia's exploration trough came later than the U.S. and Canadian troughs, which reflected the fact that Australia experienced no upturn in exploration following the mineral price increases of 1973 and particularly 1974. This evidence, therefore, is consistent with the argument that policies adversely influenced exploration.

Nevertheless, more than half of the drop in exploration expenditures occurred in fiscal year 1971–72 (July 1–June 30), whereas the Labour Party did not assume power until December 1972. Much of the decline occurred before Labour Party rule, and Labour policies put in place, therefore, cannot have been directly responsible for most of the decline. Having said this, however, there may be some merit to the view that expectations or uncertainty over the future course of policy discouraged exploration in Australia relative to other countries.

We cannot therefore determine precisely how important changing policies and risks were in shaping Australian exploration trends. But it is likely that other factors also were important. In the search for these other factors, it is necessary to consider Australian exploration over a longer period.

During the late 1950s and early 1960s, Australia's mineral potential was reassessed following significant discoveries of bauxite, manganese, lead, zinc, and particularly iron ore. Of twenty-nine important discoveries made in Australia between 1956 and 1965, twelve were iron ore deposits (King, 1973). The initial search for iron ore was stimulated by the emergence of the Japanese steel industry and its ore requirements. Following the first discoveries, Australia lifted its ban on iron ore exports, which provided additional incentives for exploration. Other important discoveries included the Weipa bauxite deposit, the Macarthur and Woodcutters lead-zinc deposits, and the Groote Eylandt manganese deposit.

Although these discoveries may have led to a steady reappraisal of Australia's geologic potential in the early 1960s, several discoveries in 1966 apparently sparked the dramatic surge in exploration and discovery that lasted until the early 1970s. Generally rising mineral prices during the late 1960s and 1970 merely added fuel to the fire. The most notable discoveries in 1966 were the Kambalda nickel region in Western Australia and several phosphate deposits in Queensland. Of fifty-eight mineral discoveries made between 1966 and 1973 (King, 1973), among the most notable were the Woodlawn lead-zinc deposit in New South Wales, four uranium deposits in the Northern Territory (Narbarlek, Ranger, Koongarra, Jabiluka), and fourteen nickel sulfide deposits near Kambalda in an area that, as noted by Sullivan (1974), "was previously considered a gold province, and any mention that nickel might be present was generally regarded with great skepticism."

During the mid-1970s, as noted earlier, Australia experienced a deep and prolonged trough in minerals exploration. In addition to declining mineral prices and adverse government policies, the very success of exploration between 1956 and 1973 may have contributed to the slump. By 1973 the exploration boom had run its course—that is, significantly diminishing marginal returns to additional exploration had been reached. This occurred in the absence of advances in geologic concepts of mineral occurrence in Australia, improvements in exploration technology, or changes in other factors influencing prospective financial returns to exploration. Moreover, the companies were worried about marketing what had been discovered already. "So much iron ore, bauxite, and nickel had been found there was little point in funding further exploration" (Barnett, 1980).

Between 1978 and 1982, Australia experienced another exploration boom, which began about a year or so after the Liberal-National Party coalition returned to power replacing the Labour Party. Although some of the restrictive Labour Party policies were eased—most notably the restrictions on foreign-owned equity—mineral prices (reflecting overall mineral market conditions) and geology once again became the important driving forces shaping Australian exploration trends. Prices for many minerals, including gold, rose in 1979 and 1980, encouraging exploration not only in Australia but around the world. Australia's geology is particularly suited to gold deposits. This, along with dramatically higher gold prices during the late 1970s, helped create "gold rush" conditions there. Large expenditures for uranium, diamonds, and coal also helped to raise total exploration expenditures.

In summation, government policies in Australia were perhaps responsible for the relatively deep and long trough in exploration expenditures compared with other countries in the early and mid-1970s. These policies would have discouraged exploration in Australia by raising the costs and risks of exploring and mining there. Still, much of the decline in exploration expenditures occurred before the Labour Party assumed power. Weakening mineral markets (reflected in generally falling mineral prices) and diminishing returns to additional exploration due to the large number of recent mineral discoveries added to the downturn. More generally, changing mineral prices largely explain the two exploration peaks because of their effects on expectations about future prices and profitability and on the availability and cost of exploration financial capital.

Canada

Rumblings also were heard in Canada during the 1970s about the effects of government policies—primarily federal and provincial tax law modifications—on exploration. Canada showed large exploration peaks in 1970 and 1980–81 but only a small peak in 1975 corresponding to the high 1974 mineral prices.

DeYoung (1977, 1978) studied the effects of Canadian tax law changes in the early 1970s on mineral exploration in that country, and his results provide some support for the belief that policies and political risks are of considerable importance to exploration. By comparing trends over time in mineral exploration in various Canadian provinces with neighboring regions, he concluded that the major effect of tax changes was to shift the location of exploration from one political jurisdiction to another.

The most striking example is British Columbia, where exploration expenditures, drilling footage, and claim-staking declined in 1974 and 1975 following provincial tax changes. These included two layers of royalties that were viewed as particularly onerous by mining companies. At the same time, exploration appears to have increased in the neighboring U.S. states of Alaska, Washington, and Montana and in the Yukon and Northwest Territories of Canada. These territories, under federal control, were not significantly influenced by federal-provincial tax battles (DeYoung, 1978). At the international level, DeYoung (1977) noted that Canadian mining companies are estimated to have spent 60 percent of their exploration funds abroad in 1975 compared with 20 percent in 1971. Nevertheless, DeYoung admitted that it is clearly impossible to separate the impact of changing tax laws from other factors, such as mineral prices and geologic concepts, that affect exploration budget allocations.

The United States

A number of changes in U.S. government policies during the last fifteen years, particularly in land management regulation, are often cited as having made exploration and mining there less attractive relative to other areas. Concern over federal land management appears to have been especially high during two periods. The first period occurred in the mid-1970s when the withdrawal of public lands from mineral development attracted much attention. In a much-quoted article, Bennethum and Lee (1975) of the U.S. Department of the Interior estimated that the percentage of public lands withdrawn from locatable mineral development increased from 17 percent in 1968 to 53 percent in 1974.[1] The second period occurred in the late 1970s when concern centered around the closing of some Alaskan lands to mineral development. This followed several years in which Alaska was a favorite frontier area for many exploration groups (see, for example, Mullins, Lawrence, and Deschamps, 1977; and Hodge and Oldham, 1979). Many explorationists were disturbed by what they viewed as misguided policy that discouraged exploration in geologically attractive areas. Nevertheless, as demonstrated below, it is far from clear that U.S. federal and state policies have been on balance more onerous or restrictive than policies in other countries.

Table 2–1 shows how the U.S. share of overall exploration expenditures for nine companies has changed since 1970. Table 2–2 displays U.S. share data for a broader sample of U.S. and Canadian companies over a shorter period. The data reveal no common and distinctive

Table 2–1. U.S. Share of Total Mineral Exploration Expenditures by Nine Companies, 1971–82
(percentage)

				U.S. mining companies				U.S. oil companies	
Year	Amax	Asarco	Noranda	1	2	3	4	1	2
1971	n.a.	n.a.	13	n.a.	n.a.	34	n.a.	48	90
1972	54	>50	16	n.a.	38	22	49	34	71
1973	45	ca. 50	17	80	13	18	49	30	53
1974	64	n.a.	26	83	19	29	53	40	57
1975	68	>63	31	n.a.	21	31	49	36	55
1976	54	n.a.	34	100	35	35	55	35	66
1977	48	n.a.	29	87	47	38	68	33	66
1978	57	>60	33	62	30	42	67	30	60
1979	58	>75	39	73	29	46	62	27	57
1980	58	ca. 75	36	58	20	48	56	22	47
1981	n.a.	ca. 75	38	65	28	51	54	29	40
1982	n.a.	n.a.	33	76	34	27	49	29	37

Note: n.a. = not available. ca. = circa.
Sources: Annual reports of Amax, Asarco, Noranda; personal communications from the unnamed U.S. mining and oil companies that require anonymity.

[1]This sort of estimate has been inherently imprecise because of incomplete data. In a separate estimate, the Office of Technology Assessment (1979) concluded that in 1975 some 34 percent of federal onshore land was formally closed to hardrock mineral development and that access to an additional 6 percent was highly restricted. The remaining 60 percent was subject to only moderate or slight restrictions.

trends that would allow definitive conclusions to be drawn. For oil company #1, the U.S. share declined sharply in the early 1970s as the company initiated exploration efforts in Canada and Australia; the share fell gradually from 1974 to 1980 and then rose in 1981. At Amax the large increase in the U.S. share in 1974 probably resulted from greatly expanded exploration for energy minerals in the United States; in the late 1970s the share increased mostly because of large expenditures on detailed exploration and evaluation at the Mt. Emmons (Colorado) and Mt. Tolman (Washington) molybdenum prospects. Asarco's share of expenditures devoted to U.S. exploration appear to have increased gradually during the last ten years, although this inference is based on incomplete data. The data in table 2–2 are more comprehensive and indicate no declining U.S. share by U.S. companies, but they cover a period far too brief for assessing the effects on exploration of policies over any extended period of time.

Foreign exploration groups, on the other hand, have argued that U.S. policies on balance are favorable for exploration compared with policies elsewhere. According to Cook (1983, p. 12), "the entry of foreign exploration companies into the U.S. will continue and possibly accelerate. . . . The attraction of the U.S. for exploration is the political stability and generous foreign investment policies." Exploration expenditures by Noranda in the United States rose from 13 percent of

Table 2–2. U.S. Share of Total Mineral Exploration Expenditures by U.S. and Canadian Companies, 1980–84
(absolute figures in thousands of current US$)

	U.S. companies			Canadian companies		
	Expenditures		U.S. share	Expenditures		U.S. share
Year	Total	In U.S.	(percent)	Total	In U.S.	(percent)
1980	513,063	316,408	62	191,007	9,442	5
1981	604,788	364,560	60	222,548	13,856	6
1982	520,173	302,613	58	163,137	8,858	5
1983	421,912	276,571	66	148,048	21,714	15
1984	309,133	206,417	67	160,261	10,145	6

Note: Based on fifty-nine U.S. companies in 1980–82 (representing an estimated 80 to 90 percent of total mineral exploration expenditures by U.S. companies), sixty-two companies in 1983 (an estimated 90 percent of total expenditures), and forty-eight companies in 1984 (80–90 percent). Based on thirty-nine Canadian companies in 1980–82, fifty-two Canadian companies in 1983, and fifty-eight Canadian companies in 1984 (60–70 percent). Includes expenditures for base and precious metals, other metals, uranium, and industrial minerals.

Sources: G. A. Barber and S. Muessig, "Minerals Exploration Statistics for the Years 1980, 1981, and 1982," *Economic Geology* vol. 79, no. 7 (1984) pp. 1768–1776; Barber and Muessig, "1983 Minerals Exploration Statistics: United States and Canadian Companies," *Economic Geology* vol. 80, no. 7 (1985) pp. 2060–2066; and Barber and Muessig, "1984 Minerals Exploration Statistics: United States and Canadian Companies," *Economic Geology*, in press.

total expenditures in 1971 to 39 percent in 1979 and 33 percent in 1982. For a larger number of Canadian companies, exploration expenditures in the United States more than doubled between 1982 and 1983, rising from 5 to 15 percent of their total exploration expenditures. In 1984, however, they fell sharply (table 2–2). Finally, Consolidated Gold Fields, based in London, spent more than 80 percent of its direct or corporate exploration (excluding expenditures by subsidiaries and partially owned companies such as Newmont) in North America, much of it in the United States (Kramer, 1983). Nevertheless, these shifts toward the United States could just as well be due to other factors that offset any adverse effects on policy.

The implications, based on very fragmentary data and incomplete analysis, are that there is no statistical evidence to support the view that on balance U.S. government policies toward explorationists are any more onerous than those pursued by other governments. Indeed the data, although admittedly weak, would even allow one to conclude that U.S. policies may be more favorable than most.

South Africa[2]

Four distinct periods of mineral exploration have occurred in South Africa since 1960. During the 1960s expenditures remained relatively constant, with precious metals exploration accounting for about 60 percent of exploration outlays.

Between 1969 and 1974, total real expenditures rose from 6.9 to 27.6 million rands (constant 1975 prices), due almost solely to increased exploration for base and alloy metals—including copper, lead, zinc, tin, nickel, and molybdenum—from 1.5 million rands to 22.1 million rands. Rising prices in 1970, 1973, and 1974 as well as the widespread mineral shortages and fears of long-term supply problems explain part of this increase. But a reappraisal of South Africa's geologic potential for copper, lead, zinc, and nickel following several significant discoveries by foreign companies is likely to have been the primary determinant. The most important discoveries were Palabora by Newmont and Rio Tinto Zinc, Prieska by Anglovaal and U.S. Steel, Black Mountain and Broken Hill by Phelps Dodge, and Gamsberg by Newmont. Beukes argues that these and other foreign multinational firms were attracted to South Africa during this period at least in part because of the changing political climate in southern Africa following the gaining of independence by former European colonies. Exploration apparently was redirected away from some newly independent states because of perceived political risks.

[2]This section is based largely on Beukes (forthcoming 1987).

Thus, as noted at the beginning of the chapter, South Africa experienced no peak in exploration expenditures in the early 1970s in contrast to exploration peaks that occurred in some other countries: expenditures, particularly for nonprecious metals, increased steadily from 1970 to 1977. This can be explained partly by changing mineral prices, partly by ore deposit discoveries and associated changes in perceptions of geologic potential, and partly by the political climate in southern Africa.

In the third period, between 1975 and 1979, expenditures for base and alloy metals fell by more than 50 percent, while expenditures for uranium, precious metals, and coal increased somewhat. These changes correlate roughly with relative prices for these minerals. During the most recent period, 1979–83, precious metals regained their position as the most important exploration target, again due largely to changes in relative mineral prices. All this may change if gold and silver prices decline and political risks are reappraised in light of increasing domestic turmoil and its worldwide impact.

Developing Countries

A widely held belief is that exploration by major corporations has declined markedly in developing countries in recent years as a result of ill-conceived host government policies and associated uncertainties regarding the potential for additional changes in the future. This has led to fears of future mineral supply problems because current exploration is not being undertaken in the most geologically attractive areas (in developing countries). These countries, the argument goes, should become increasingly important mineral producers as the higher-grade deposits of developed mining countries are depleted.

Data from a group of European mining companies frequently are used to support these concerns. As shown in figure 2–4, the developing country share of total exploration expenditures for these companies fell from about 40 percent in the late 1960s to 20 percent and less in the early 1970s, with the steepest decline occurring in 1970 (from 41 to 21 percent). Since then the developing-country share has fluctuated between 8 and 22 percent. How supportable is the inference that 1970 represents a watershed year in perceptions of political risk?

Resource nationalism in mineral-rich countries had been simmering since the early 1960s and apparently reached a peak in 1970 in its effect on exploration. For example, in Mexico the policy of "Mexicanization" instituted in 1961 required that "all new mining ventures including exploration had to be 51 percent Mexican" (Walthier, 1976), even though operating mines had twenty years to comply. In 1964

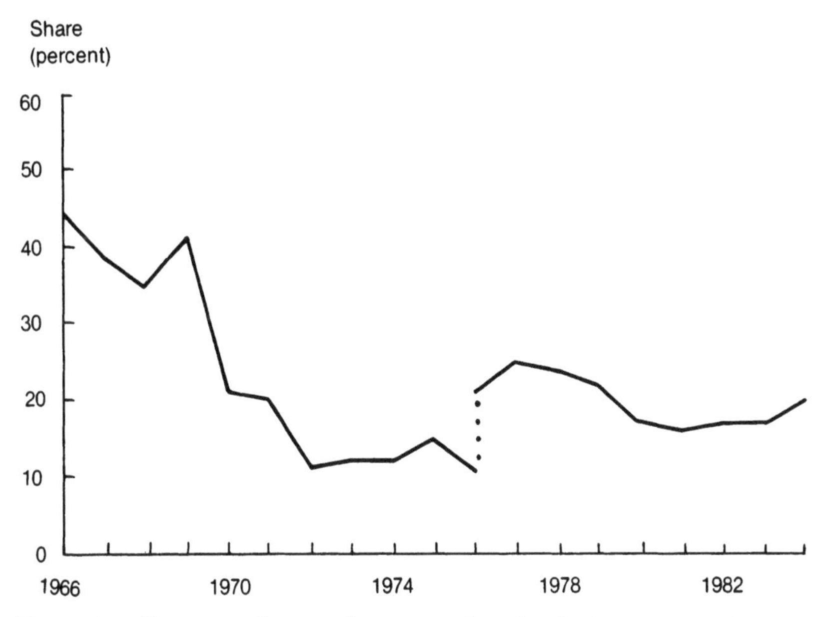

Figure 2–4. European Community companies: developing country share of exploration expenditures, 1966–84. The discontinuity in 1976 is the result of a change in the composition of the European companies from which data were collected. *Source:* Based on P.C.F. Crowson, "A Perspective on Worldwide Exploration for Minerals," in John E. Tilton, Roderick C. Eggert, and Hans H. Landsberg, eds., *World Mineral Exploration: Trends and Economic Issues* (Washington, D.C., Resources for the Future, forthcoming 1987) table 2-A-9.

President Eduardo Frei of Chile called for nationalization of foreign-owned copper mines; this occurred following the election of President Salvador Allende in 1969 (the nationalizations were not completed until 1971; Moran, 1973). In Africa in the late 1960s Zambia acquired controlling interest in its two largest mining groups, Anglo-American Corporation and Roan Selection Trust, and Zaire nationalized Union Minière's holdings in its country.

Thus one might expect the large drop in shares in 1970 to be reflected in expenditures in these areas, namely Latin America and Africa (excluding South Africa). However, the share of exploration expenditures in African developing countries was too small to be responsible for much of the decline, even though in relative terms the African share fell more than 35 percent from 5.1 percent in 1969 to 3.2 percent in 1970. The Latin American share simply remained negligible at 0.5 percent. The large decline in the developing-country share, noted above, reflects primarily a decline in Oceania, excluding Australia, from 17.0 percent in 1969 to zero percent in 1970. Radetzki

(1982) maintains that if the Bougainville copper project in Papua New Guinea is excluded from the statistics from 1966 to 1969, the decline in developing-country share is much less dramatic—to 21 percent from about 30 percent rather than from more than 40 percent. Thus a large part of the decline in 1970 reflects the movement of one project from the exploration to development stage and not necessarily increased perceptions of political risk. (This instance, incidentally, reveals the hazards inherent in drawing inferences from aggregate data.)

Still there may be some validity to the allegedly adverse role of government policies and political risks. Since 1970 the developing-country share of the European companies has been consistently lower than it was during the late 1960s. Although this trend cannot be attributed solely to one factor, the resource nationalism of some developing countries in earlier years must be considered a plausible contributing factor. Adverse policy changes toward mining companies and the specter of possible future changes can do nothing but raise the costs and risks and thus lower potential returns for a firm considering an investment in such a country, all other factors remaining the same.

As for the fear that future mineral supplies from developing countries will be inadequate because of inadequate current exploration efforts in those countries, the limited amount of data is clearly insufficient for making such an inference. The European data represent only part of overall exploration expenditures. North American and other multinational companies conduct significant exploration efforts in developing countries; for one U.S. oil company, the share of its exploration expenditures devoted to developing countries increased from 2.3 percent to 10.2 percent between 1973 and 1982 (Eggert, forthcoming 1987). Furthermore, even if the developing-country share of exploration expenditures by multinational companies has declined, exploration by other organizations, such as indigenous state-owned mining companies and private domestic firms, is substantial and may have partially or wholly replaced multinational exploration.

In Brazil, for example, only US$45 million of the estimated US$132 million, or 34 percent, spent on mineral exploration (metallic and nonmetallic minerals and coal) in 1983 came from international mining companies; the remaining 66 percent came from Brazilian concerns (Crowson, forthcoming 1987). In Mexico the major private companies in the Chamber of Mines, the government, and smaller domestic companies are estimated to have spent in total an average of approximately US$50 million on mineral exploration in the early 1980s (Crowson, forthcoming 1987). According to Johnson and Clark (forthcoming 1987) considering all sources of funds (public and private, domestic and foreign) mineral exploration expenditures in Botswana averaged $11.2 million from 1970 to 1982 ($13.3 million for 1980–82), and in Papua New Guinea they averaged $8.3 million from 1970 to 1982 ($6.9 million for 1980–82, expressed in constant 1982 U.S. dollars).

More generally these data indicate the problems of discussing "developing countries" as a single, homogeneous lot. A danger exists of obscuring more than is revealed. A great deal of diversity exists among countries usually included in the developing country category. Some countries, such as Chile, Malaysia, and Zambia, have long and rich histories of mining and exploration, while others have little or no mining tradition. Some countries have well-developed infrastructures and basic geologic and geophysical information that are readily available, while others are infrastructure- and information-poor. Some countries have policies to encourage mineral exploration by foreign and domestic organizations, while others do not. Given these and other differences among developing countries, is it reasonable to expect exploration to have occurred similarly everywhere in the developing world? It makes infinitely more sense to discuss exploration in specific countries for particular minerals.

Summing Up

The foregoing analysis indicates that, for two reasons, mineral prices strongly influence year-to-year changes in mineral exploration expenditures and thus explain many of the similarities among trends in particular countries. First, changing mineral prices, as a composite indicator of changes in mineral supply and demand conditions, are an important determinant of expectations about future prices and profits from exploration and mining; in other words, expectations about the future are shaped largely by the present and recent past. Second, mineral prices influence mining revenues and, in turn, the availability and cost of capital for financing exploration by private, as opposed to public, organizations. For both reasons exploration expenditures and mineral prices generally tend to rise and fall together, with expenditures lagging prices by about a year.

To explain the differences among exploration expenditure trends in particular countries it is necessary to call on country-specific factors, the two most important of which are geologic potential and government policies/political risks. The geologic potential for ore discovery varies widely among countries at any point in time, and varies over time for particular countries as the understanding of ore deposit geology improves, discoveries are made, and prices and costs change for particular minerals or deposit types. Changes in government policies and political risks, on the other hand, alter the investment attractiveness of a country by affecting the costs and risks of exploring and mining in that country.

APPENDIX 2–A. MINERAL EXPLORATION IN THE SOVIET UNION[1]

The discussion in chapters 1 and 2 centered on mineral exploration in the so-called market-economy countries, which excludes an important part of the world, the centrally planned economies of Eastern Europe, China, and the USSR. Although relatively little is known in the Western world about exploration in these areas, a recent paper by Astakhov, Denisov, and Pavlov (forthcoming 1987) sheds some light on exploration in the Soviet Union. It examines, among other topics, the organization of prospecting and exploration activities, recent trends in the level and distribution of expenditures, and planning and decisionmaking in the USSR.

This appendix—drawing solely on the paper by Astakhov, Denisov, and Pavlov for factual information on Soviet exploration (unless noted otherwise)—describes exploration in the USSR and compares it with exploration in the market-economy countries.

The sequence of exploration activities in the USSR is, not surprisingly, similar to the sequence of activities in Australia, Canada, the United States, and other countries in the Western world. Soviet exploration is officially broken into six sequential stages, which are distinguished by their goals, scales of activity, and search methods. Regional geologic and geophysical surveying, the first stage, is followed, in turn, by prospecting, preliminary exploration, and detailed exploration. During these four stages, the focus of the search is continually narrowed, beginning with surface examination of large areas (during prospecting) and leading to detailed, subsurface investigation of particular mineral occurrences (during detailed exploration). The final two Soviet stages occur near or at operating mines; exploration in the vicinity of an operating mine establishes new reserves in the immediate area, while operational exploration attempts to better delineate or extend known reserves. Moreover, Soviet explorationists generally rely on the same techniques as their Western counterparts: geologic mapping, geochemical sampling, geophysical surveying, and drilling to name the most important.

Despite these similarities between exploration in the USSR and elsewhere, large differences can be found in other aspects such as organization and decisionmaking. The USSR Ministry of Geology is responsible for about 85 percent of Soviet prospecting and exploration jobs, whereas in the United States and other Western countries a much more diverse set of institutions is involved in mineral exploration. The Western government's role typically is restricted to providing regional geologic and geophysical information necessary for reconnaissance and target selection; the private sector is largely responsible for actual exploration.

Exploration planning and decisionmaking in the USSR are importantly influenced by the Soviet system of central planning. In the first of four mineral development steps, long-term scenarios (supply/demand matrices) for the national economy are developed. These scenarios serve as external inputs into the second step, which calculates the necessary mining and, in turn, the

[1]Based largely on Eggert (1985b).

optimal level and distribution of exploration expenditures to meet projected mineral requirements. In the third step, the Ministry of Geology allocates funds among the various regions and national organizations responsible for exploration, based on the estimates of the previous step. Lastly, regional and local organizations allocate funds among their projects.

Despite what appears to be a top-down and almost deterministic system, apparently substantial negotiation and give-and-take between the hierarchical levels occurs as the central plan is developed each year. This permits, for example, demand forecasts to be modified in light of new cost conditions resulting from mineral discoveries or new cost-saving technologies. Still Soviet exploration is heavily influenced by mine and metal production goals that are imposed on the mineral sector of the economy by the central plan, whereas exploration in the West is controlled to a much larger degree by free-market forces.

Annual Soviet expenditures for hard-rock mineral exploration more than doubled between the periods 1961–65 and 1976–80 (measured per five-year period), or an average annual increase of about 5 to 6 percent. They were scheduled to increase by some 6 percent per year between 1981 and 1985. Although some of this increase represents expanded exploration work (for example, more geologic mapping, geochemical sampling, or drilling), some also represents increased exploration costs associated with working in more remote areas and difficult geologic environments.

The distribution of Soviet expenditures for mineral exploration also has changed since the early 1960s. Prospecting, the second stage of Soviet exploration (roughly comparable to reconnaissance in the terminology of chapter 1), accounted for more than 50 percent of total exploration expenditures during the five-year period 1976–80, compared to some 40 percent during 1961–65. The increase is due in part to shifts in activity to more remote, eastern regions of the Soviet Union, where there is insufficient geologic information to conduct more detailed investigations. As for the distribution of prospecting expenditures among particular minerals, between 1961–65 and 1976–80 expenditures increased by more than 200 percent for antimony, tin, and apatite; by close to 100 percent for iron ore, lead, and zinc; and by about 50 percent for manganese and bauxite. Expenditures rose only 20 to 30 percent for copper and mercury, while those for molybdenum, titanium, and potassium held steady or declined.

three

THE EPISODIC NATURE OF EXPLORATION FOR PARTICULAR MINERALS

Chapter 2 examined mineral exploration over time in particular countries. We now move on to examine another aspect, that of mineral or deposit-type sought. This chapter compares and contrasts exploration trends for iron ore, bauxite, copper, molybdenum, uranium, and gold. The chapter's main thesis, drawn from Rose and Eggert (forthcoming 1987), is that exploration and discovery of particular minerals or deposit types are generally episodic over time.

An idealized episode of exploration and discovery begins with increased incentives for exploration for a certain mineral or deposit type. Incentives take several forms, any one of which alone can catalyze the episode. Rising prices and demand often encourage increased estimates of future revenues from exploration and subsequent mining. New or revised geologic models—idealized sets of physical and chemical characteristics common to most deposits of a particular type that guide reconnaissance exploration and follow-up work—alter the way explorationists search for mineral deposits and reduce exploration costs and geologic risk. Improved exploration and production technologies reduce exploration, mining, or processing costs associated with a particular type of deposit. All forms of incentives make it potentially more profitable than before to discover an ore deposit.

The episode continues with stepped up exploration and some discoveries, either in areas where little had been known previously about the geologic potential for ore occurrence or in areas with known mineralization or mining. Initial discoveries provide additional incentives for exploration over the short term by reducing the geologic risks associated with an area—that is, the odds of making additional discoveries increase because not only is the region generally favorable geologically, but the favorable characteristics are actually associated with ore. A bandwagon effect is often apparent: A successful leader or innovator is followed by a host of imitators.

Over the longer term, however, discoveries of a particular mineral or deposit type often discourage further exploration by greatly expanding actual or potential sources of supply. Falling or stagnant prices and demand also may lead to the final stage of the idealized episode: declining exploration and discovery.

The following examples of exploration demonstrate that, despite the foregoing generalization, many types of cycles exist. Some episodes, for example, continue to be broad and intense for a decade or more, while others come and go in only a few years. There may be little or no regularity among episodes for a particular mineral; long periods of little or no exploration may occur between episodes. Also the relative importance of the various driving forces, such as mineral prices or technological change, tends to vary from episode to episode and from mineral to mineral.

Iron Ore and Bauxite[1]

Iron ore and bauxite both enjoyed a similar period of large-scale worldwide exploration from the late 1940s to the early 1970s. Prior to this period, most of the iron ore used in the United States had been mined in Minnesota and Michigan. Most iron ore used in Europe had been mined in Sweden, France, and Great Britain, and most of the bauxite in France and Hungary. During and just following World War II, however, a fear arose that traditional sources of North American and European ore were being depleted at rates that would lead to physical shortages and higher prices in the near future, particularly as demand rose for aluminum and iron and steel products used to reconstruct Europe and to feed economic growth in the United States.

As a result North American and European companies became the leading participants in the surge of immediate postwar exploration, and they searched farther away from traditional steel and aluminum producing areas than in the past. U.S. companies searched actively in South America and Mexico for iron ore and in the Caribbean region, (especially Jamaica) for bauxite. European companies explored actively in western Africa. By the mid–1950s, a number of discoveries had been made, many in areas where mineralization had been known for a long time but that had been deemed too remote to warrant detailed exploration and evaluation. The airborne magnetometer made it relatively simple to detect several large magnetic iron ore bodies in Canada.

[1]This section is based on Eggert (1985a).

From the mid-1950s to the early 1970s, iron ore and bauxite exploration continued at high levels in many parts of the world. But compared with earlier postwar exploration, the incentives then were slightly different. Much of the growth in iron ore and bauxite demand occurred in Japan and western Europe rather than in the United States. Late in the period, some developing countries sought to produce steel or aluminum and as a consequence increased their demand for iron ore or bauxite. Substantial exploration programs continued into the early 1970s, encouraged by forecasts of substantial growth in demand for steel and aluminum products and, in turn, iron ore and bauxite. The most notable discoveries between the mid-1950s and the early 1980s were made in Australia, where seventeen iron ore and eight bauxite deposits were found (King, 1973). Subsequent production from these discoveries that significantly affected world iron ore supplies were made in Brazil, Canada, and Venezuela. Important discoveries of bauxite were made in Brazil, Guinea, and Jamaica.

Fears of depletion and rising demand were not the only stimulants for the increased geographic spread of iron ore and bauxite exploration. The unit costs of long-distance ocean transport for bulk materials fell substantially in the 1950s and 1960s, permitting explorers to look in remote areas; reductions of 50 to 60 percent in transport costs were not uncommon (Manners, 1971). Economies of scale in mining and processing also encouraged the search for large, high-grade deposits in some remote areas rather than for smaller, lower-grade deposits near operating mines. At the same time, however, developments in concentrating, sintering, and pelletization stimulated delineation of large, low-grade iron occurrences, called taconites, particularly in the United States. Taconites became the preferred ores of many U.S. steelmakers—they were readily available from domestic sources, and iron pellets, processed from the taconite, worked very efficiently in the blast furnace. Consequently exploration in the United States for high-grade natural iron ores virtually ceased.

Technological change also influenced bauxite exploration. Just as advances in processing techniques allowed taconites to be profitably mined, so too did technologic progress in alumina refining enable bauxites with a wider range of chemical compositions to be used. Before the early 1950s, the specifications of the Bayer process for refining bauxite (Al_2O_3 + H_2O + impurities) into alumina (Al_2O_3) were strict—bauxites had to contain more than 55 percent alumina and less than 8 percent Fe_2O_3 and 2–3 percent silica. Since then modifications of the Bayer process have enabled the use of an almost entirely new class of bauxites that contain as little as 30 percent available alumina and as much as 28 percent Fe_2O_3. These lower-grade bauxites now account for about two-thirds of production in the Western world. They are the aluminous laterites, which occur in tropical regions of Australia, Africa, South America, and Asia. Technologic advances in processing, therefore, have stimulated exploration by the major aluminum companies for these lateritic deposits since the mid-1950s.

From the mid-1970s to the mid-1980s, very little iron ore and bauxite exploration occurred. The little that was done consisted largely of detailed evaluations or extensions of known deposits, undertaken with an eye toward development, particularly in Australia, Brazil, and western Africa. Lower projections of future growth in steel and aluminum consumption, coupled with the successful exploration of the 1950s and 1960s, created an apparently prolonged overabundance of identified iron ore and bauxite resources. Moreover, the poor financial performance of many steel and, to a lesser extent, aluminum companies reduced the availability of internal funding for exploration. Tables 3–1 and 3–2 show that the share of iron ore of total metals exploration activity from the mid-1970s to the mid-1980s was small in the two countries where data were readily available. In Australia iron ore's share of expenditures fell from 12.6 to 4.3 percent between 1976/77 and 1984/85, while in the United States the share of iron ore exploration drilling footage was below 1 percent between 1974 and 1982.

Table 3–1. Metallic Mineral Exploration Expenditures in Australia, 1976/77–1984/85
(million A$, current prices)

Fiscal year (July 1–June 30)	Private exploration expenditures: total metallic[a]	Iron ore	Iron ore's share of total (percent)
1976/77	111.97	14.09	12.6
1977/78	125.30	10.66	8.5
1978/79	138.02	8.13	5.9
1979/80	200.83	10.50	5.2
1980/81	340.00	14.42	4.2
1981/82	397.04	15.64	3.9
1982/83	317.67	10.20	3.2
1983/84	350.29	9.32	2.7
1984/85	365.40	15.80	4.3

Sources: Based on P. C. F. Crowson, "A Perspective on Worldwide Exploration for Minerals," in John E. Tilton, Roderick G. Eggert, and Hans H. Landsberg, eds., *World Mineral Exploration: Trends and Economic Issues* (Washington, D.C., Resources for the Future, forthcoming 1987); and Australian Bureau of Statistics, *Mineral Exploration Australia (1984–1986)*, Catalogue no. 8407.0.

[a]Includes copper, lead, zinc, silver, nickel, cobalt, gold, iron ore, mineral sands, tin, tungsten, uranium, other metals.

Table 3–2. Exploration Drilling in the United States, 1973–82
(thousands of feet by all methods)

Year	Total metallic[a]	Iron ore	Iron ore's share of total (percent)
1973	19,489	239	1.2
1974	19,278	159	0.8
1975	24,088	125	0.5
1976	13,634	91	0.7
1977	20,900	85	0.4
1978	23,039	58	0.3
1979	18,244	72	0.4
1980	15,450	39	0.3
1981	10,800	58	0.5
1982	6,140	30	0.5

Sources: Based on P. C. F. Crowson, "A Perspective on Worldwide Exploration for Minerals," (statistical appendix) in John E. Tilton, Roderick G. Eggert, and Hans H. Landsberg, eds., *World Mineral Exploration: Trends and Economic Issues* (Washington, D.C., Resources for the Future, forthcoming 1987); and U.S. Bureau of Mines, *Minerals Yearbook* (Washington, D.C., Government Printing Office, 1974–83).

[a]Includes copper, gold, iron ore, lead, molybdenum, silver, tungsten, uranium, zinc, other metals.

Copper

Porphyry copper exploration went through a long period of activity from the late 1940s to the middle 1970s.[2] Table 3–3 lists twenty-seven porphyry deposits discovered in the United States between 1949 and 1980, including eighteen that have produced copper. Elsewhere around the world, deposits were discovered in Argentina, Canada, Chile, Ecuador, Iran, Panama, Papua New Guinea, Peru, the Philippines, and Yugoslavia, among other countries. Steadily growing copper demand and improvements in a geologic model of ore occurrence (making exploration easier and less costly) were incentives for porphyry copper exploration, and initial discoveries in each area stimulated additional exploration. In addition over the last thirty-five years discoveries have greatly expanded known and potential sources of copper. Thus discoveries—along with stagnating copper demand in many industrialized countries and increased capital costs for large, low-grade porphyry deposits relative to smaller, higher-grade deposits—have contributed to the declining interest of explorationists in porphyry copper deposits between the mid-1970s and 1980s.

[2]Porphyry copper deposits are a geologic class of mineral deposits in which the copper (typically grading 0.5 to 2.0 weight percent) is relatively uniformly distributed throughout a granitic igneous rock (termed a "porphyry").

Table 3–3. U.S. Porphyry Copper Deposits Discovered Between 1945 and 1984

Discovery year	Deposit	Company
1949	Copper Cities, Arizona*	Miami
1950	Silver Bell, Arizona*	Asarco
1951	Yerington, Nevada*	Anaconda
1951	Pima-Mission, Arizona*	Pima-Asarco
1955	Esperanza, Arizona*	Duval
1957	San Xavier, Arizona*	Asarco
1958	Tyrone, New Mexico*	Phelps Dodge
1958	Safford (KCC), Arizona	Kennecott
1959	Palo Verde, Arizona*	Banner
1960	Mineral Park, Arizona*	Duval
1961	Safford (PD), Arizona	Phelps Dodge
1962	Sierrita, Arizona*	Duval
1963	Sacaton, Arizona*	Asarco
1964	Twin Buttes, Arizona*	Banner
1965	Bluebird, Arizona*	Ranchers
1965	Kalamazoo, Arizona*	Quintana
1968	Sanchez, Arizona	Inspiration
1970	Helvetia, Arizona	Banner-Anaconda
1970	Copper Creek, Arizona	Newmont
1970	Red Mountain, Arizona	Kerr McGee
1970	Lakeshore, Arizona*	Hecla
1970	Florence, Arizona	Conoco
1970	Metcalf, Arizona*	Phelps Dodge
1973	Pinto Valley, Arizona*	Cities Service
1973	Copper Basin, Arizona	Phelps Dodge
1976	Casa Grande, Arizona	Hanna-Getty
1980	Hillsboro, New Mexico*	Quintana

Note: Deposits marked with asterisks have produced.
Source: Based on Arthur W. Rose and Roderick G. Eggert, "Exploration in the United States," in John E. Tilton, Roderick G. Eggert, and Hans H. Landsberg, eds., *World Mineral Exploration: Trends and Economic Issues* (Washington, D.C., Resources for the Future, forthcoming 1987).

As discussed by Rose and Eggert (forthcoming 1987), this recent episode of U.S. copper exploration is only part of a longer episodic history of copper discovery in the United States, which is displayed in figure 3–1. In the 1840s copper was discovered in northern Michigan. Only simple crushing and gravity separation were required to obtain the copper because of the ore's mineralogical simplicity; the ore was native copper rather than sulfide or oxide ore. The next major discoveries were not made until 1870 in the Butte area of Montana. The early ores at Butte were very high grade, containing 5 to 20 percent copper as well as appreciable silver and gold, which permitted underground mining.

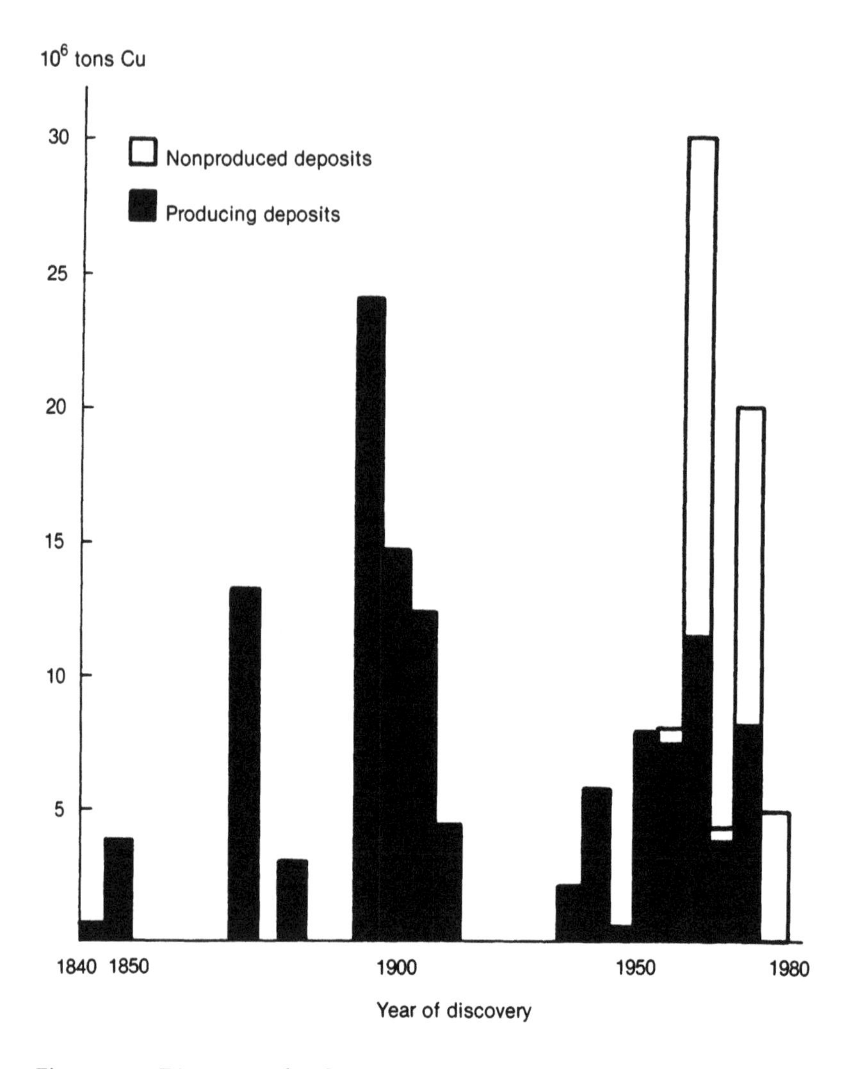

Figure 3–1. Discovery of U.S. copper reserves, 1840–1980. *Source:* Arthur W. Rose and Roderick G. Eggert, "Exploration in the United States," in John E. Tilton, Roderick G. Eggert, and Hans H. Landsberg, eds., *World Mineral Exploration: Trends and Economic Issues* (Washington, D.C., Resources for the Future, forthcoming 1987).

The beginning of the twentieth century witnessed the birth of porphyry copper mining at Bingham, Utah. Mining engineer Daniel Jackling realized that although copper at Bingham averaged only 2 percent by weight of the rock (similar to waste rock at Butte), it was relatively evenly distributed throughout an enormous volume of granitic rock and therefore could be profitably mined using new large-scale, surface-mining methods. Jackling's recognition that this low-grade, copper-bearing rock was ore, rather than merely mineralized rock, revolutionized copper mining at a time when copper demand was rising due mainly to the rapid spread of electrification (see Prain, 1975). As a consequence nine other porphyry copper deposits (seven in the United States and two in Chile) were discovered and developed between 1900 and 1915.

Mining at Bingham and other porphyry copper deposits moved to progressively lower-grade ore by expanding the scale of operations and by adopting a new flotation process for separating copper from its host rock. (By 1925 the average ore grade was about 1 percent copper, and about 0.8 percent in 1960. By 1972, 1.7×10^9 tons of 0.71 percent copper ore remained at Bingham, compared to total past production of 1.2×10^9 tons of 0.91 percent copper; Jackling's original estimates were 1.2×10^7 tons of 2 percent copper ore! Gilmour, 1982).

Therefore technologic improvements in mining and mineral concentration rather than exploration and discovery of new ore deposits satisfied growing copper demand from 1915 to about 1950. Little copper exploration occurred until after World War II, when copper supplies relative to increasing postwar demands were perceived to be growing short. Companies responded by initiating or increasing copper exploration programs (stimulated also by government loans and price supports during the Korean War). The extended postwar episode of porphyry copper exploration, which lasted until the mid-1970s, thus had begun.

Molybdenum

Molybdenum experienced a much shorter and more recent episode of widespread exploration compared with iron ore, bauxite, and copper. During the mid- and late-1970s, molybdenum became an important target for many companies, yet between 1981 and 1986, molybdenum exploration declined precipitously.

A few exploration groups actually became interested in porphyry molybdenum deposits as far back as the 1960s. Kennecott and Amax developed geologic models then and began large-scale exploration that continued through the 1970s. In 1965 Amax discovered its Henderson ore body (Colorado), which had no surface exposure, by studying the surface and near-surface geology, predicting where ore mineralization should occur at depth, and drilling to a depth of more than 1,000 meters. (Mine production began in the early 1980s.)

Nevertheless, porphyry molybdenum exploration did not become widespread until the mid-1970s, when the factors encouraging exploration included rising molybdenum prices, reflecting tighter metal supplies relative to demand. Spot prices (unadjusted for inflation) rose from $2.90/lb. in 1975 to $9.22/lb. in 1979 and $22.00/lb. in 1980, and then fell to less than $4.00/lb. by 1983. Prices in mid-1986 stood at about $3.00/lb. Rising prices presumably encouraged anticipation of higher future prices and profitability from molybdenum mining and, in turn, exploration. Another factor related to prices that encouraged optimism about potential profits from molybdenum exploration was the importance of the metal in many uses of energy production—including conventional oil wells, enhanced recovery techniques, arctic oil and gas production, and solar panels—at a time when energy exploration was booming. At least seven deposits were discovered in the United States from 1975 through 1981 (table 3–4).

The precipitous decline in the share of exploration expenditures for molybdenum in the first half of the 1980s is largely the result of the then current and potentially chronic state of excess mine capacity created by new mines and recent discoveries. Between 1960 and 1980, two mines (Climax and Questa, discovered in about 1915 and 1957, respectively) satisfied U.S. demand and allowed for significant exports. Since then three new mines have opened at Henderson (Colorado), Tonopah (Hall, Nevada), and Thompson Creek (Idaho); and numerous additional deposits, all currently on indefinite hold, are in the offing, including Quartz Hill (Alaska), Mount Hope (Nevada), Mount Emmons (Colorado), Mount Tolman (Washington), and Pine Grove (Utah). Adding to the perception of oversupply is the potential for significantly more by-product production of molybdenum from many porphyry copper mines.

Table 3–4. U.S. Porphyry Molybdenum Deposits Discovered Between 1940 and 1984

Discovery year	Deposit	Company
1957	Questa, New Mexico*	Molycorp
1962?	Hall, Nevada*	Anaconda
1965	Henderson, Colorado*	AMAX
1975	Thompson Creek, Idaho*	Cyprus
1976	Quartz Hill, Alaska	U.S. Borax
1977	Pine Grove, Utah	Phelps Dodge
1977	Mt. Emmons, Colorado	AMAX
1978	Nye County, Nevada	UV Industries
1978	Mt. Tolman, Washington	AMAX
1981	Mt. Hope, Nevada	Exxon

Note: Deposits marked with asterisks have produced.

Source: Based on Arthur W. Rose and Roderick G. Eggert, "Exploration in the United States," in John E. Tilton, Roderick G. Eggert, and Hans H. Landsberg, eds., *World Mineral Exploration: Trends and Economic Issues* (Washington, D.C., Resources for the Future, forthcoming 1987).

Uranium

Worldwide exploration for uranium has gone through two distinct episodes since World War II. The first, from the late 1940s to the late 1950s, was associated with the military procurement programs of the United States, the United Kingdom, and France. The second episode was stimulated by commercial investments in nuclear power and extended from the late 1960s to the early 1980s, experiencing a dramatic surge in exploration from about 1974 to 1980.[3]

During the late 1940s and 1950s the United States—and to a lesser extent the United Kingdom and France—developed and expanded their arsenals of nuclear weapons, thereby increasing demand for uranium and, in turn, for uranium exploration. Exploration in various parts of the world—particularly Canada, the United States, Australia, France, and parts of Africa—was further encouraged by price guarantees, tax favors, and even discovery bonuses. The Australian government, for example, paid out bonuses of A$50,000 to the discoverers of the Rum Jungle and Mary Kathleen deposits, discovered in 1949 and 1954, respectively. Developed with financing from the United States and United Kingdom, these two deposits, along with the Radium Hill deposit discovered in the 1940s, accounted for most of Australia's uranium production and exports in the 1950s.

In Canada uranium exploration also was encouraged by financial incentives from the U.S. and U.K. weapons procurement programs, but exploration occurred on a larger scale than in Australia (Neff, 1984). Although exploration in Canada began in 1942 with government-sponsored programs, it increased substantially in the late 1940s and early 1950s after a ban on private exploration was lifted. By 1959 Canada had twenty-three uranium mines and nineteen treatment plants in five districts (OECD/IAEA, 1982). As for the United States, early postwar exploration was sponsored by the U.S. Atomic Energy Commission (AEC). During the 1950s the private sector assumed a greater role in exploration, and government exploration was phased out. Exploration activity, particularly by private firms, concentrated on areas of known mineralization such as the Colorado Plateau, the Wyoming Basins, and the Gulf Coastal Plain (OECD/IAEA, 1982).

The U.S. government was a much larger purchaser of uranium than either the British or French governments. Most of the uranium purchased by the AEC came from the United States and Canada. The United States also imported uranium from the Belgian Congo (now Zaire), particularly in the late 1940s and early 1950s, and later from

[3]A thorough and succinct historical summary of events in the international uranium market, on which this section draws extensively, is Neff (1984).

Australia, South Africa, and Portugal. The United Kingdom relied primarily on South Africa and Australia for its uranium procurements, while France obtained imports from Gabon and Niger, in addition to mining uranium domestically.

By about 1960 the market for nuclear weapons had become saturated, a fact that was signaled in 1959 when the AEC announced it would not extend its uranium purchase contracts with Canadian producers past 1962 and 1963, although purchases eventually were stretched out until 1967. Western world production (that is, outside the centrally planned economies) fell from a high of 32,052 metric tons of uranium (MTUs) in 1960 to 14,939 MTUs in 1967, a decline of some 53 percent, with peak-to-trough production declines over similar periods of 48 percent in the United States, 54 percent in France and its Francophone African associates (including Gabon), 77 percent in Canada, and 78 percent in Australia. As a consequence of this collapsing uranium market, due in part to the success of exploration in the 1950s, very little uranium exploration occurred between 1960 and the mid- to late-1960s, particularly in the two countries hardest hit by production cutbacks, Australia and Canada.

Incentives for the second episode of uranium exploration, lasting from the late 1960s to the early 1980s, were fundamentally different from those of the first period. During this episode exploration was encouraged by the first civilian uses of uranium in nuclear power generation and by expectations for large growth in this market rather than by uranium demands of weapons-procurement programs. Uranium exploration in many parts of the world increased substantially in the latter half of the 1970s in response to wildly higher uranium prices and the tangled web of changes in underlying supply and demand conditions they reflected. These changes included most importantly (1) increased current and potential future uranium demand as a result of new reactor orders and the belief that nuclear power would play a major role in electric power generation over the next several decades and (2) rapid change and uncertainty internationally due to several factors. These included western Europe's drive to break U.S. dominance in uranium enrichment, unexpected delays in uranium mining in Australia, interruption of uranium exports from Canada following India's explosion of a nuclear device with plutonium produced in a research reactor from Canada that was to be used for peaceful purposes, France's halt of uranium sales, forward sales by Westinghouse Corporation of uranium for which it had no contracts with primary producers, lifting of the U.S. ban on processing imported uranium for domestic use, a uranium producers cartel, and changes in U.S. enrichment policy requiring fixed rather than flexible commitments for enrichment of fuel core up to eight years in advance (Neff, 1984).

The second episode of uranium exploration came to a close in the early 1980s as uranium prices tumbled. Several factors played a part in the price decline: uranium production capacity had doubled in little more than five years; uranium reserves and resources had increased dramatically as a result of new discoveries (particularly in Australia and Canada); and projections of nuclear power growth were lowered as expected electricity demand leveled off, nuclear regulatory restrictions grew, and prospective costs of nuclear power rose. These factors resulted in an apparently chronic oversupply of uranium and a virtual halt to exploration.

In contrast with the first episode of uranium exploration, this second episode can be well documented. Figure 3–2 displays uranium exploration expenditures between 1966 and 1984 by a group of European companies and in the United States. Figure 3–3 gives expenditure estimates over a shorter period, 1972–83, for selected other countries and the world (excluding the centrally planned economies). The most striking characteristic of nearly all the expenditure trends is a pronounced peak in the late 1970s. Worldwide expenditures are estimated to have increased by nearly 600 percent between 1972 and 1979. The parallel between changing uranium prices (figure 3–4) and exploration expenditures is also noteworthy. Rising prices, reflecting

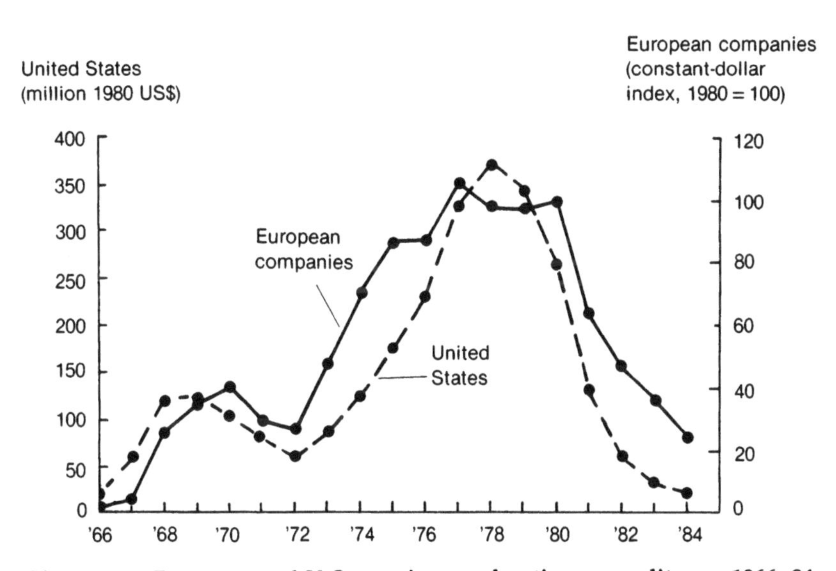

Figure 3–2. European and U.S. uranium exploration expenditures, 1966–84. The data for Europe are expenditures *by* European companies, normalized to account for a change in 1976 in the composition of firms included in the survey. The U.S. data are expenditures *in* the United States and are from the U.S. Department of Energy. *Source:* appendix table A–3.

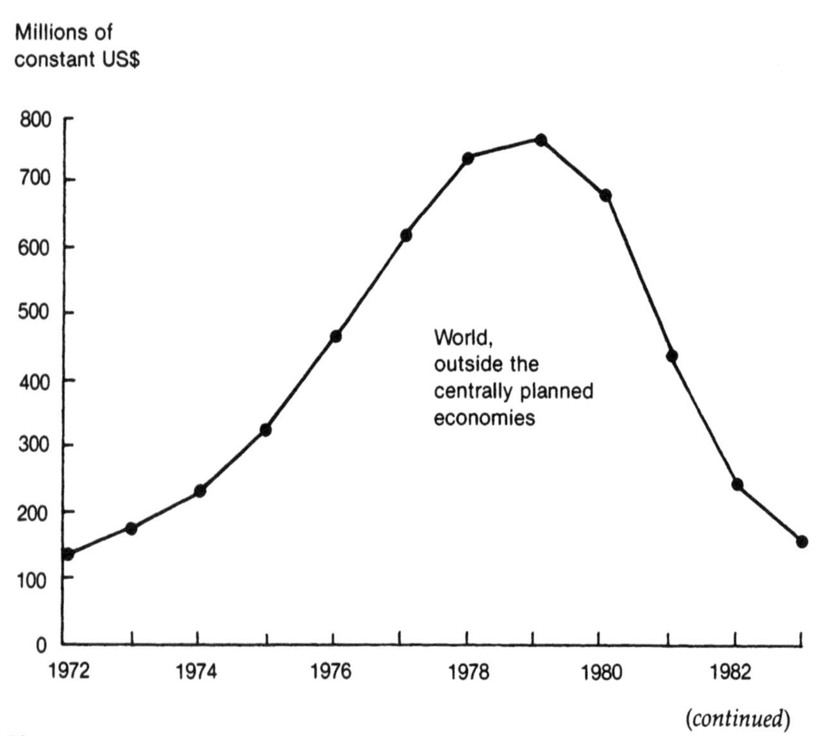

(continued)

Figure 3–3. OECD-IAEA estimates of uranium exploration expenditures, 1972–83. *Source:* appendix table A–4.

Figure 3–3. *(continued)*

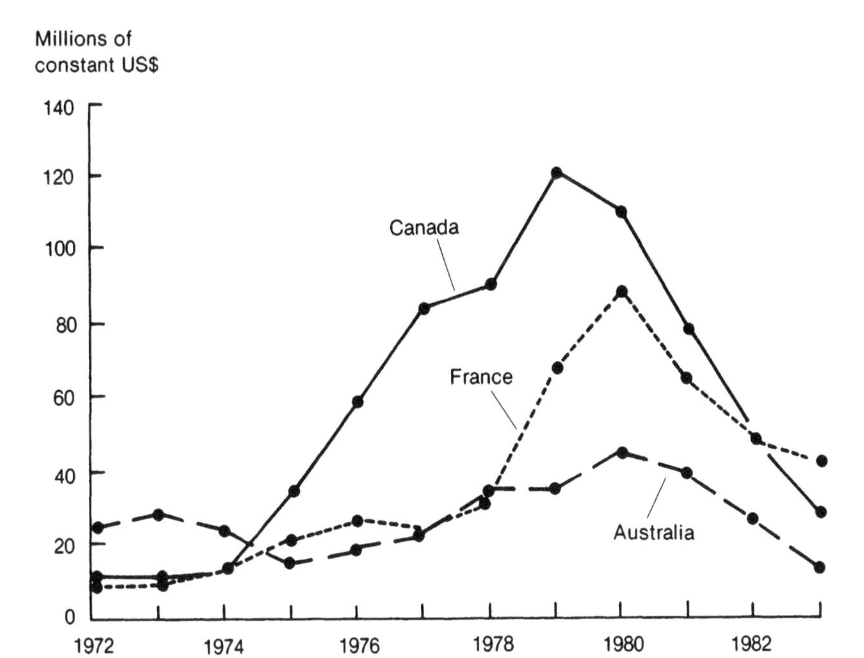

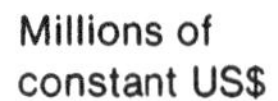

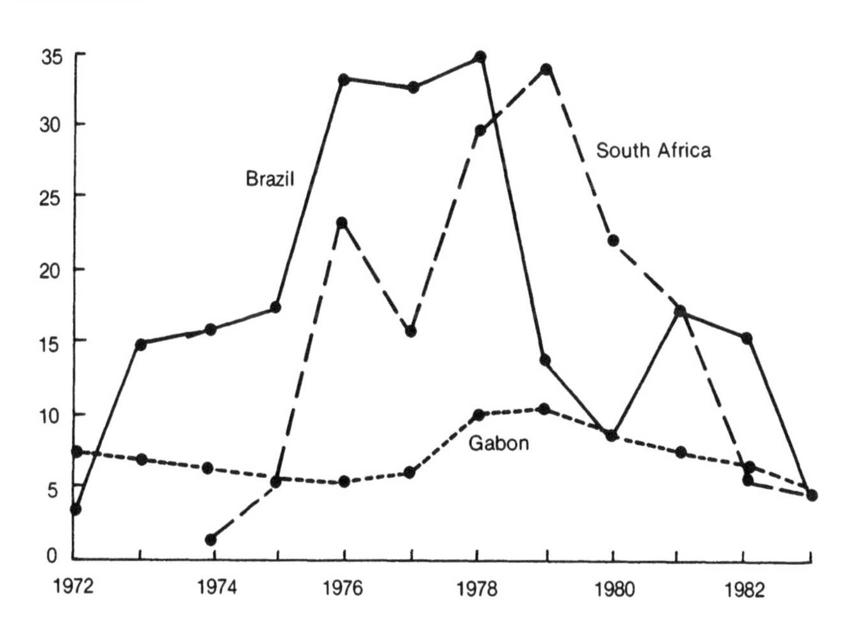

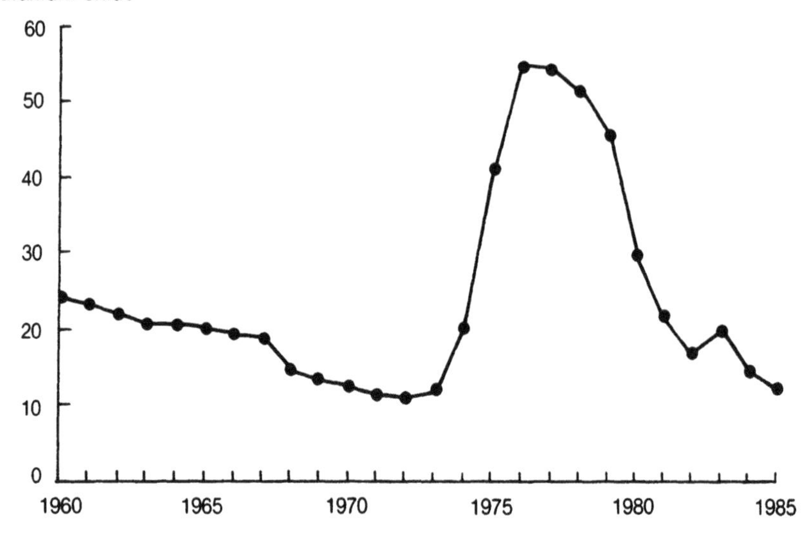

Figure 3–4. Uranium prices, 1960–85. *Source:* appendix table A–2.

tighter current and potential future supplies with respect to demand, stimulated exploration by encouraging expectations of continued high prices and profitability from uranium mining. Exploration expenditures tended to lag price changes by a year or so because it took time to respond to changing incentives.

The United States and Canada together accounted for some 66 percent of cumulative Western world expenditures for uranium exploration between 1972 and 1983 (table 3–5). France and Australia made up 9 percent and 7 percent, respectively, while the rest of the noncommunist world accounted for the remaining 18 percent. France, however, was the most intensively explored country by a wide margin during the period (table 3–5). Table 3–6 shows that estimates of exploration expenditures for a particular country can vary substantially from source to source even though they purport to measure the same activity; although the OECD/IAEA and DOE data generally agree between 1972 and 1975, the estimates from 1976 to 1983 differ by as much as 70 million U.S. dollars.

How successful was exploration during the second episode of uranium exploration, and how did the success vary across countries? Table 3–7 allows for a simple comparison. So-called reasonably assured resources (equivalent to reserves) more than doubled and estimated additional resources (roughly comparable to the U.S. "probable potential resources" category) nearly tripled between 1967 and 1983. The largest absolute and relative increases occurred in Australia. In

Table 3–5. Uranium Exploration: Cumulative Expenditures and Annual Average Intensity in Selected Countries, 1972–83

Country	Cumulative exploration expenditures (million 1983 US$)	Share of world total (%)	Average annual expenditures (million 1983 US$)	Area (square miles)	Intensity average annual ($/sq. mi.)
Australia	388.5	6.5	32.38	2,966,200	10.92
Brazil	251.7	4.2	20.98	3,286,470	6.38
Canada	827.3	13.8	68.94	3,849,670	17.91
France	536.2	8.9	44.68	210,040	212.74
Gabon	103.5	1.7	8.63	102,317	84.30
India[a]	51.9	0.9	6.49	1,269,420	5.11
West Germany	98.2	1.6	8.18	96,011	85.20
Mexico[b]	28.1	0.5	3.12	761,604	4.10
South Africa[c]	191.3	3.2	19.13	435,868	43.89
United States	3117.8	51.9	259.82	3,623,420	71.70 (61.34)[e]
31 other countries[d]	411.3	6.8	n.a.	n.a.	n.a
Total	6005.8	100.0			

Note: n.a. = not applicable.
Source: appendix table A–4.
[a]Expenditures cover 1972–79 only.
[b]Expenditures cover 1972–80 only.
[c]Expenditures cover 1974–83 only.
[d]Outside the centrally planned economies.
[e]Calculated with expenditure data from the U.S. Department of Energy rather than OECD/IAEA. See table 3–6

Table 3–6. Two Estimates of Uranium Exploration Expenditures in the United States, 1972–84
(million US$, current prices)

Year	OECD/IAEA[a]	DOE[b]
1972	32.4	32.4
1973	49.5	49.5
1974	80.9	79.1
1975	130.2	122.0
1976	195.9	170.7
1977	293.5	258.1
1978	371.5	314.3
1979	385.6	315.9
1980	332.6	267.0
1981	180.3	144.8
1982	97.8	73.6
1983	56.7	38.9
1984	n.a.	26.5

Note: n.a. = not available.
Source: appendix table A–3.
[a]Organisation for Economic Co-operation and Development and the International Atomic Energy Agency.
[b]U.S. Department of Energy.

Table 3–7. Estimates of Reasonably Assured and Estimated Additional Reserves of Uranium, 1967 and 1983
(1,000 MTU)

	Reasonably assured[a]		Estimated additional[b]	
Country	1967	1983	1967	1983
Australia	8	314	2	369
Canada	154	176	223	360
France	35	56	15	27
Gabon	3	19	3	1
South Africa[c]	158	191	12	99
United States	139	131	250	501
Other	38	581	35	207
Total	535	1,468	540	1,564

Note: The 1967 data represent uranium that could be produced for a price of less than US$10/lb. U_3O_8 (1967 $), whereas the 1983 data are for material available at a forward cost of less than US$30/lb. U_3O_8 (1983 $). The categories are roughly comparable over time (see Neff, 1984).
Sources: OECD/IAEA as reported in Thomas Neff, *The International Uranium Market* (Cambridge, Mass., Ballinger Publishing Company, 1984, Appendix B); and OECD Nuclear Energy Agency and the International Atomic Energy Agency, *Uranium Resources, Production and Demand* (Paris, OECD, 1984).
[a]Uranium in known deposits for which tonnage and grade have been calculated based on direct observations and which could be mined and processed with current technology.
[b]Additional uranium suggested by direct geologic evidence, generally associated with known deposits and well-defined geologic trends. Estimates in this category have a higher margin of error.
[c]The 1967 South Africa figures include Namibia, whereas the 1983 numbers do not.

Canada although reasonably assured reserves barely changed, estimated additional resources grew substantially. In the United States reasonably assured reserves actually fell while estimated additional resources doubled.

Gold

Gold is in the midst of an episode of sizable exploration that began in the mid-1970s. As shown in figures 3–5 and 3–6, gold exploration has increased substantially in absolute terms (expenditures) and as a share of total metals exploration. Between 1970 and 1983, real exploration expenditures for gold grew more than seven times in the United States and more than three times in both Canada and South Africa. By 1983 in the United States, Canada, and Australia, gold exploration accounted for between 44 and 65 percent of total mineral exploration expenditures, compared with 15 to 35 percent in 1980 and 14 to 25 percent in 1975 (excluding Australia, for which 1975 data are not available). Also by 1983 gold was close to regaining its position of dominance in South African exploration: up to 44 percent of expenditures from 12 percent in 1974, compared with more than 50 percent for nearly all the 1960s.

What accounts for the recent interest in gold exploration? Gold prices, undoubtedly, have been the most important factor. Between 1934 and 1968 the nominal price of gold was controlled at US$35 per ounce and thus declined in real terms. Since then gold's price has been allowed to fluctuate freely according to market conditions and has risen considerably (figure 3–7). By 1985 gold's price was more than four times higher in real terms than in 1970 (nearly ten times in nominal terms), even though it had fallen by more than half since 1980. Substantially higher prices transformed into ore what previously had been nothing more than mineralized rock. Higher gold prices have created enormous expectations of potential revenues from gold mining, compared with expectations fifteen years ago, and thus have stimulated exploration. Much of recent gold exploration has occurred in historical areas of gold mining.

In addition to higher gold prices, three other factors have contributed to the recent episode of gold exploration. First, recent improvements in heap leaching and carbon-in-pulp recovery of gold have reduced production costs at many of the low-grade deposits typical of recent discoveries (Brown, 1983). Second, in a more limited sense the goal of diversifying production among a number of minerals has contributed to the importance of gold exploration. Companies have sought to reduce swings in earnings caused by price instability through

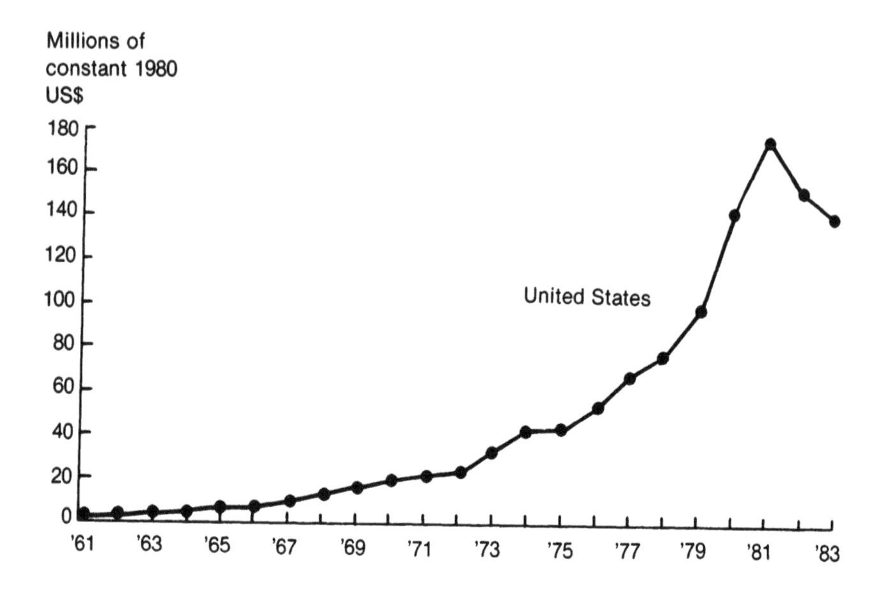

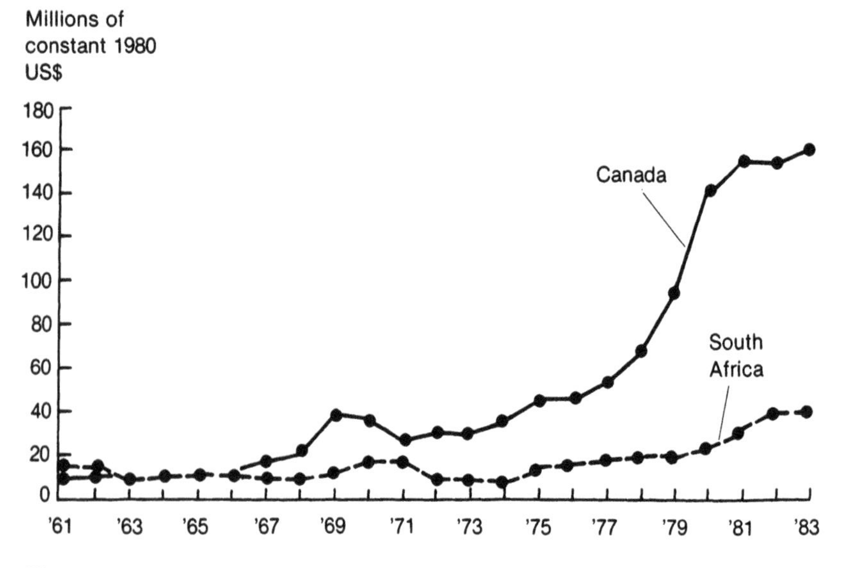

Figure 3–5. Gold exploration expenditures in selected countries, 1961–83.
Source: appendix table A–5.

diversifying into several minerals. Two companies, St. Joe and Inco, for example, initiated large gold exploration programs as part of diversification activities. Third, advances in exploration technology and geologic science have improved the ability of explorationists to locate gold mineralization. In the field of geochemistry, explorationists now can reliably measure chemical elements associated with gold in the parts per billion (ppb) range. Explorationists also have benefited from

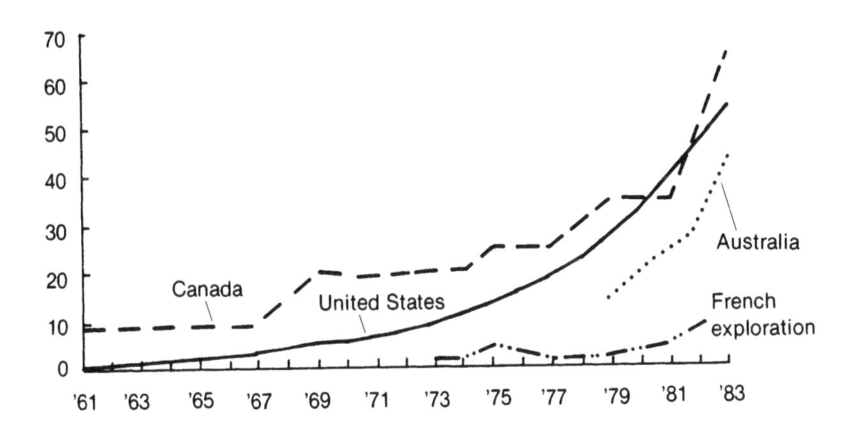

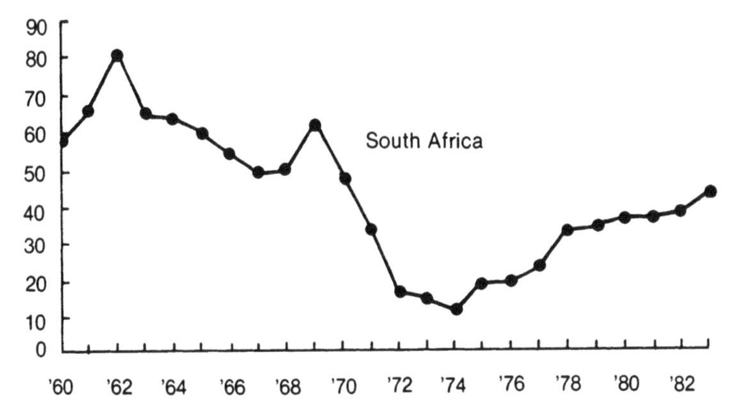

Figure 3–6. Gold's share of exploration expenditures in selected countries and by French explorationists, 1961–83. *Source:* appendix table A–5.

improved geologic models of gold occurrence in sedimentary rocks and in hot spring (volcanic) environments (Paul Bailly, former president of Occidental Minerals, personal communication).

The recent emphasis on gold exploration is reflected in the timing of major gold discoveries in the United States since 1940 (table 3–8). Of twenty-one discoveries only two—Carlin and Cortez—were made in the 1960s and eight in the 1970s, whereas the remaining eleven were made between 1980 and 1982. Although particular years of discovery may be open to some question (that is, discovery dates may be off by a year or two), the broad trend is undeniable: For the last ten years or so, we have witnessed dramatic growth in gold exploration and discovery.

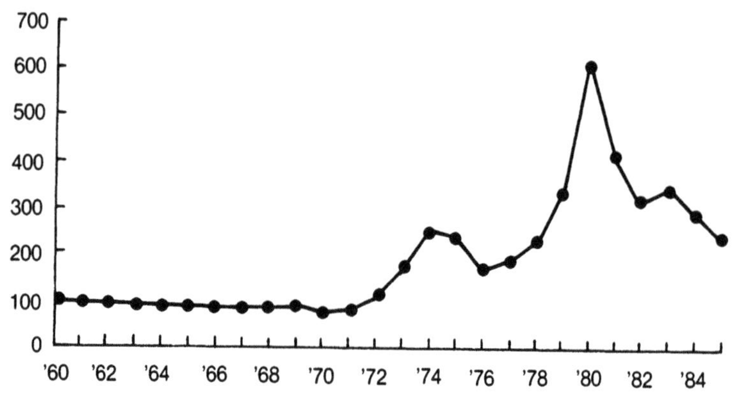

Figure 3–7. Gold prices, 1960–85. *Source:* appendix table A–2.

Table 3–8. Major Gold Deposits Discovered in the United States Between 1940 and 1982

Discovery year	Deposit	Company
1962	Carlin, Nevada*	Newmont
1966	Cortez, Nevada*	Amex
1971	Pinson, Nevada*	Cordex-Rayrock
1974	Cornucopia, Nevada*	Standard Silver
1974	Round Mountain, Nevada*	Copper Range
1975	Candelaria, Nevada*	Occidental
1976	Jerritt Canyon, Nevada*	FMC-Freeport
1976	Ortiz, New Mexico*	Goldfields
1978	Kearsarge, California	Picklands-Mather
1978	Alligator Ridge, Nevada*	Amselco
1980	Maggie Creek, Nevada*	Newmont
1980	McLaughlin, California*	Homestake
1980	West End, Idaho*	Superior Oil
1980	Borealis, Nevada*	HIMCO
1981	Mercur, Utah*	Getty Oil
1981	Golden Sunlight, Montana*	Placer Amex
1982	Gold Quarry, Nevada*	Newmont
1982	Rain, Nevada	Newmont
1982	Zaca, California	California Silver
1982	Boulder Creek (Dee), Nevada*	Rayrock
1982	Horse Canyon, Nevada*	Placer Amex

Note: Deposits marked with asterisks have produced.
Source: Based on Arthur W. Rose and Roderick G. Eggert, "Exploration in the United States," in John E. Tilton, Roderick G. Eggert, and Hans H. Landsberg, eds., *World Mineral Exploration: Trends and Economic Issues* (Washington, D.C., Resources for the Future, forthcoming 1987).

four

ECONOMIC PRODUCTIVITY OF MINERAL EXPLORATION

Chapters 2 and 3 focused on the inputs to mineral exploration, namely, the level of expenditures devoted to particular geographic regions and mineral target types. This chapter examines the outputs of exploration in relation to these inputs, or in other words, exploration productivity. It reviews the work of a number of researchers on mineral exploration productivity in Australia, Canada, France, and the United States. In doing so the chapter addresses such questions as: Has it become more expensive over time to discover economic mineral deposits as the easier-to-find deposits generally have been discovered first? If so, what are the implications for future mineral supply? Has productivity varied across countries, and why? But first a word is in order about what productivity means and how it can be measured.

Although productivity can be measured in a variety of ways, all measurements relate outputs of a process or activity with associated inputs. Some measures are ratios of physical output and input units, such as tons of raw steel produced per worker-hour. Others use economic units, such as value of steel production divided by the wage rate. Still others use both physical and economic units—for example, value of steel production per worker-hour.

Calculating mineral exploration productivity, therefore, is a matter of assigning physical or economic values to exploration output (economic mineral deposits) and input (exploration activity). Perhaps the simplest measure is the number of deposits discovered per year. Simplicity is about the only virtue of this measure; it says nothing about the size or potential worth of a discovery (the variance among deposit qualities is large), and it ignores the amount of effort required for discovery.

A second productivity measure assigns economic units to exploration inputs, while again simply counting the number of discoveries over a given period as a proxy for exploration output. The result is the number of deposits discovered per exploration dollar, or, more commonly, the inverse, which is the cost per discovery.

A third measure of productivity assigns economic units to exploration output: gross value of mineral discoveries over time, calculated by multiplying metal content of annual discoveries by appropriate metal prices. This measure distinguishes among various deposit sizes and thus is an improvement over the first measure, although it too ignores the amount of effort required for a discovery.

A fourth productivity calculation assigns economic values to both outputs and inputs: gross discovery values/exploration expenditures, or success ratios.

All four types of productivity measures have weaknesses from the viewpoint of an economist studying the role of exploration in mineral supply. First of all exploration output is valued in gross rather than net terms, if it is valued at all. The value of exploration to society depends not only on the number of discoveries or gross value of contained metal but also on the capital and operating costs of producing metal. Some discoveries are better than others. For example, a high-grade discovery in an established mining area should be assigned a higher value than low-grade, remote discovery (other factors such as the amount of recoverable metal being the same) because mining and infrastructure costs per pound of metal will be higher at the second discovery. The second weakness is that none of the four productivity measures accounts for the time value of money. A productive output from exploration benefits companies and society only several years after exploration costs are incurred. It is misleading to directly compare money spent today with future revenues without considering the financial opportunities foregone in the intervening period. A dollar today is worth more than a dollar tomorrow because at the very least the same money could earn the going interest rate in a savings account.

One way around these weaknesses is to calculate productivity as the discounted net returns from exploration. The net return from exploration is the difference between exploration benefits (value of outputs) and costs (value of inputs). Exploration benefits, in turn, are the net financial returns from mineral deposits that are discovered and then developed and mined; in other words, they are the gross revenues from the sale of refined metal minus the costs of development, mining, and mineral processing (including a reasonable rate of return on these postexploration activities). Exploration costs are those directly associated with exploration such as geologist salaries, transportation, drilling, and land acquisition (including all expenditures on unsuccessful projects).

Discounting is the process used to account for the time value of money. It is completely different from adjusting dollar values from current to constant terms, the purpose of which is to account for the diminished purchasing power of the dollar due to inflation. Discounting requires converting economic values from various time periods to a common base year, illustrated by the following equation:

$$\text{Present value of past exploration investments}_n = (B_0 - C_0)(1 + r)^n + (B_1 - C_1)(1 + r)^{n-1} + (B_2 - C_2)(1 + r)^{n-2} \ldots + (B_{n-1} - C_{n-1})(1 + r) + (B_n - C_n)$$

The present value of past exploration investment in year n is the sum of discounted net returns, $B_i - C_i$, from years 0 through n (B_i and C_i are exploration benefits and costs, respectively, in year i). The discount rate, r, represents a company's or a society's time preference for money. (Calculating the present value of past investments is actually a matter of inflating rather than discounting because yesterday's dollar was worth more than today's.) To incorporate expected future returns from exploration, an analogous calculation can be made by estimating future benefits, costs, and discount rates:

$$\text{Present value of future investments}_0 = (B_0 - C_0) + \frac{(B_1 - C_1)}{(1 + r)} + \frac{(B_2 - C_2)}{(1 + r)^2} \ldots + \frac{(B_n - C_n)}{(1 + r)^n}$$

Choosing an appropriate discount rate is crucial to calculating productivity in this manner. A high rate, for example, implies a strong time preference for money; past returns will be inflated relatively more, and expected future returns will be deflated relatively more than if the discount rate is lower. For a private company, the discount rate is its cost of capital and represents the weighted costs of debt and equity funds. For society as a whole, the appropriate discount rate represents a social opportunity cost or rate of return forgone by investing in exploration. Private and social discount rates may differ.

To summarize, the five types of productivity calculations introduced in this section range in complexity from simply counting the number of mineral deposits discovered annually to calculating the discounted net financial returns from exploration activity. As the calculations become more complex, so too do data requirements. Although none is conceptually ideal—the drawbacks of the fifth measure will be discussed later—each provides some information. Furthermore, they cover the range of actual exploration productivity calculations done to date.

Simple Productivity Measures

Derry and Booth (1978) tabulate the annual number of mineral deposits discovered in Canada from 1951 to 1976, and then use exploration expenditure data to calculate the annual average cost per discovery (in constant 1975 Canadian dollars). Even this apparently simple measure—counting discoveries—is not an unambiguous calculation because not all mineral deposits discovered (that is, technical or geologic successes) will actually be economic or profitable to mine. Many will not prove to be as large or potentially profitable as originally expected or hoped as more is learned about the deposit (its tonnage, grade, and mineralogy, for example) and about the costs of bringing it into production. Economic and political conditions can also change and have an effect over time. This problem of distinguishing between geologically interesting mineral occurrences and economic deposits plagues all efforts to analyze exploration productivity. The problem is particularly acute for recent exploration and discoveries because it is impossible to know whether or not many new discoveries will meet the economic and technical criteria necessary to proceed with development and mining.

Derry and Booth (1978) deal with this difficulty by defining a discovery as "a deposit of sufficient grade and tonnage to justify being brought into production under normal or average conditions of metal prices and interest on borrowed capital." The definition excludes discoveries put on indefinite hold, but it includes discoveries not developed immediately that nevertheless might be expected to be developed in a year or two if economic conditions became more favorable. Assigning a date of discovery also can be an accounting problem. Derry and Booth choose to define it as the year in which ore mineralization was first recognized (typically by drilling) rather than, for instance, the date of a geophysical or geochemical survey that identified the target area or the date of the decision to proceed with development.

Summary results appear in figures 4–1 and 4–2. The most striking feature is the large number of discoveries in the early and mid-1950s. These probably represent the first major application of airborne geophysical exploration techniques and geologic inference as part of the evolution of exploration technology from lone prospector to scientific explorationist (discussed in chapter 1).

Although the study is to be commended for stating clearly that merely counting the annual number of discoveries is unsatisfactory—because some discoveries are worth more than others—this shortcoming is so severe that the results provide little useful information for corporate or government decisionmakers. If, for instance, the study had been conducted in 1961, some researchers might have concluded

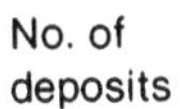

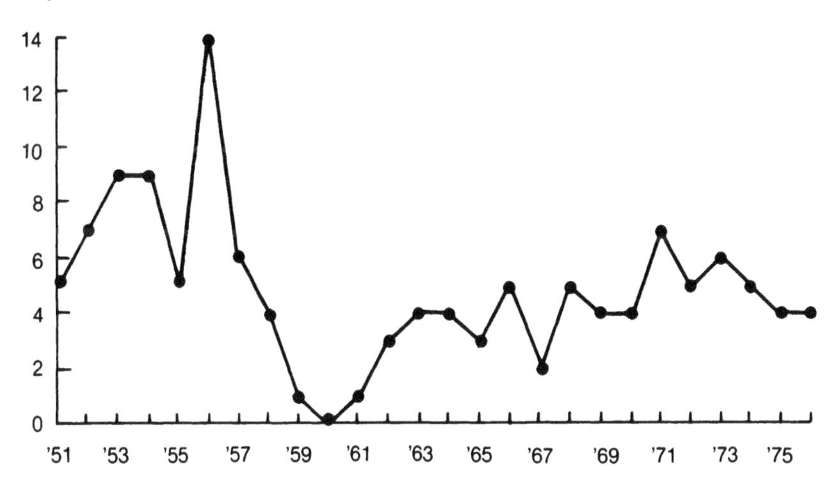

Figure 4–1. Canadian mineral deposits discovered per year, 1951–76. *Source:* Duncan R. Derry and James K.B. Booth, "Mineral Discoveries and Exploration Expenditure—A Revised Review, 1966–1976," *Mining Magazine* vol. 138, no. 5, pp. 430–433.

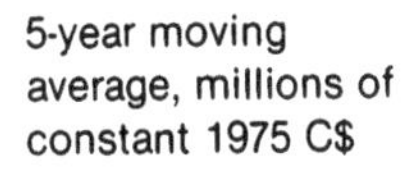

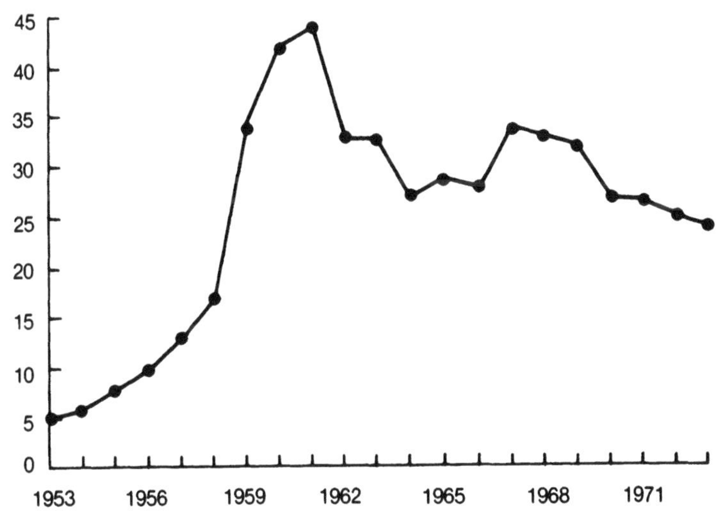

Figure 4–2. Canadian exploration costs per discovery. *Source:* adapted from Duncan R. Derry and James K.B. Booth, "Mineral Discoveries and Exploration Expenditure—A Revised Review 1966–1976," *Mining Magazine* vol. 138, no. 5, pp. 430–433.

from the pre-1961 results that mineral shortages and related supply problems would threaten mineral producers and society in ten to twenty years because of the declining number of discoveries and the increased cost of discovery. The number of discoveries fell from fourteen in 1956 to zero in 1960, and the cost per discovery (five-year moving average) increased more than four times during this short period. Yet as we well know society has survived, and mining companies now are threatened by an apparently chronic oversupply of some minerals, rather than shortages due to unsuccessful exploration.

Only part of the problem of drawing conclusions from this simple method of productivity measurement, however, is a result of merely counting the number of discoveries. More recent studies have gone beyond this simple accounting framework. Yet as the next section and the rest of the chapter demonstrate, much of the problem also results from attempting to evaluate exploration productivity in isolation from the postexploration stages of mineral supply.

Success Ratios

Studies by Cranstone (forthcoming 1987), Rose and Eggert (forthcoming 1987), and Chazan (forthcoming 1987) calculate success ratios (gross discovery values divided by exploration expenditures) for Canada, the United States, and France, respectively.

Cranstone calculates success ratios for Canadian discoveries from 1946 to 1982. The study is restricted to metallic mineral deposits, excluding lithium and iron ore. A discovery is defined as "a mineral deposit sufficiently attractive to have warranted the expenditure necessary to establish its tonnage and grade." The definition of discovery date is similar to that of Derry and Booth, mentioned previously: the year in which drilling intersected mineralization recognized, within a short period of time, to be part of an economically interesting mineral deposit.

The numerator of the success ratio, gross metal value of discoveries, is calculated by multiplying tons of metal contained in the discoveries times appropriate 1979 metal prices. Cranstone attempts to correct for systematic undervaluation of recent discoveries—because initial ore reserve estimates typically are much lower than the reserves ultimately mined—by multiplying initial ore reserves of discoveries yet to be mined by factors ranging from one to three, depending on the ore deposit type. The multipliers are based on historical records of initial and ultimate ore reserves from some 100 Canadian mines. The success ratio's denominator, exploration expenditures in 1979 Canadian dollars (C$), is derived from annual data of Canada's Energy, Mines, and Resources Department.

The ratios are displayed graphically in figure 4–3. The results are aggregated and shown in three-year periods to eliminate erratic, year-to-year fluctuations. The time trend is generally similar to Derry and Booth's time trend for number of discoveries per year: a prominent peak of successful exploration in the first half of the 1950s, representing successful use of new airborne geophysical exploration equipment and geologic inference, and since then much smaller ratios with a fairly horizontal long-term trend.

A major finding of the Cranstone study, unrelated to success ratios, confirms and quantifies the conventional wisdom that large discoveries account for the vast majority of metal discovered. Of nine hundred Canadian discoveries between 1946 and 1982, the thirty largest deposits account for about one-half of the gross value of discovered metal and the top 120 deposits about 80 percent (constant 1979 C$). On a slightly different but related tack, sixteen major mining districts (encompassing several deposits) or major deposits contain some three-quarters of the gross discovery value.

Rose and Eggert (forthcoming 1987) present success ratios for U.S. metallic mineral exploration, except for uranium and iron, from 1955 to 1983. The definitions for the discovery date and a discovery follow the definitions of Cranstone (noted earlier), except that those discoveries with gross values (tonnage times relevant 1981 mineral price) of less than US$100 million generally have been excluded. Exploration expenditure data in 1981 U.S. dollars are pieced together from various incomplete and inconsistent sources, which nevertheless provide an arguably reasonable basis for estimates of annual expenditures per five-year period. The success-ratio calculations aggregated and displayed per five-year period demonstrate that any trend over time depends critically on which discoveries are included in the calculation. When only deposits that have been mined were considered, the ratio fell by a factor of about ten between 1955–59 and 1980–83 (figure 4–4). If it is assumed, however, that some of the nonproducing discoveries will come into production the decline becomes less severe.

Chazan (forthcoming 1987) calculates success ratios for the eight most important French exploration organizations for the period 1973–82. The data are for the period as a whole—rather than year by year—and thus no trends over time are identified or discussed. Although ratios are not calculated for individual companies, they are calculated for three geographic regions of the world (French territory, the franc area, and the rest of the world) and four mineral groups (uranium, base metals, alloy and specialty metals, and precious metals). Gross

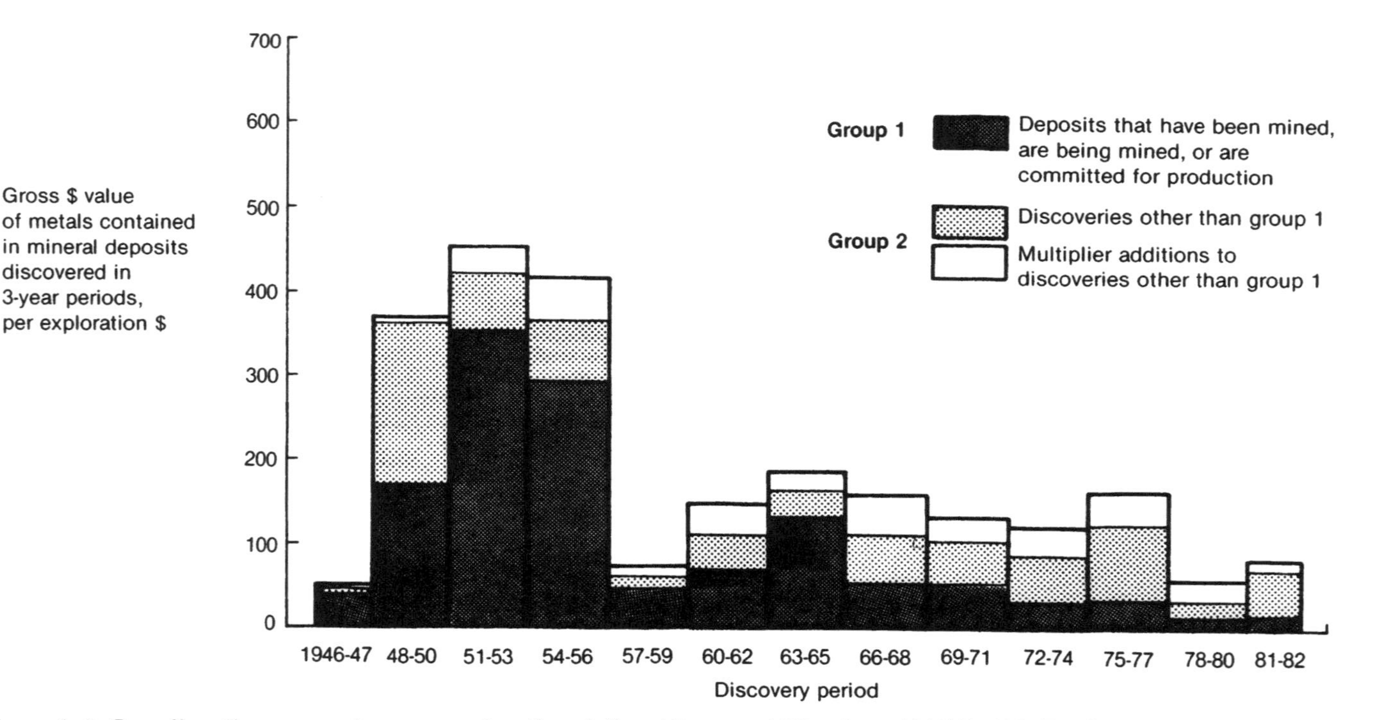

Figure 4–3. Canadian discovery values per exploration dollar at January 1979 prices, 1946/48–1981/82. *Source:* Donald A. Cranstone, "The Canadian Mineral Discovery Experience Since World War II," in John E. Tilton, Roderick G. Eggert, and Hans H. Landsberg, eds., *World Mineral Exploration: Trends and Economic Issues* (Washington, D.C., Resources for the Future, forthcoming 1987) figure 9-5.

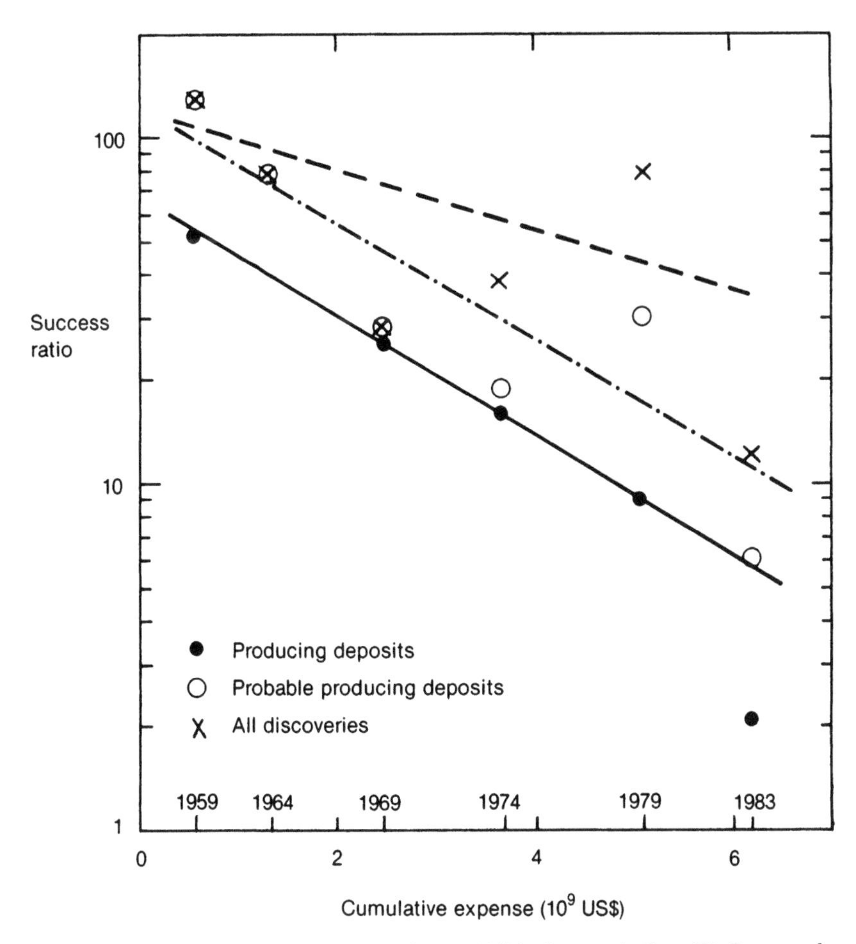

Figure 4–4. U.S. success ratios, 1955/59–1980/83. *Source:* Arthur W. Rose and Roderick G. Eggert, "Exploration in the United States," in John E. Tilton, Roderick G. Eggert, and Hans H. Landsberg, eds., *World Mineral Exploration: Trends and Economic Issues* (Washington, D.C., Resources for the Future, forthcoming 1987), figure 10-3.

discovery values are based on mineral reserves that are "unquestionably exploitable" and discovered between January 1973 and December 1982, and on 1982 international mineral prices. The reserve data, as well as exploration expenditure data in 1982 French francs, are collected by the French government in an annual survey of companies.

The success ratios are displayed in table 4–1. Regarding the geographic areas, the franc zone experienced the lowest discovery costs—that is, the highest success ratio—primarily due to uranium discoveries. As for mineral targets, precious metals exploration was markedly less successful, in terms of success ratios, than exploration for uranium, base metals, and alloy and specialty metals.

Table 4–1. Success Ratios for French Exploration, 1973–82

Minerals	Mineral targets											
	French territory			Foreign						Total		
				Franc zone			Rest of world					
	Expenditure (MF 1982)	GDV[a] (MF 1982)	Ratio[b]	Expenditure (MF 1982)	GDV[a] (MF 1982)	Ratio[b]	Expenditure (MF 1982)	GDV[a] (MF 1982)	Ratio[b]	Expenditure (MF 1982)	GDV[a] (MF 1982)	Ratio[b]
Uranium (7 companies)	1,522	40,495	27	453	41,591	92	2,004	9,118	5	3,979	91,204	23
Base metals (5 companies)	824	1,789	2	177	2,671	15	1,004	56,021	56	2,005	60,481	30
Alloy and specialty metals (7 companies)	345	9,450	27	22	—	—	56	—	—	423	9,450	22
Precious metals (5 companies)	157	794	5	82	700	19	42	300	7	281	1,794	6
Total (8 companies)	3,443	54,152	16	871	44,962	52	3,287	66,139	20	7,601	165,253	22

Notes: Principal companies only. MF = million francs. Totals are greater than the sum of the parts.

Source: Based on a draft table in Willy Chazan, "French Mineral Exploration, 1973–82," in John E. Tilton, Roderick G. Eggert, and Hans H. Landsberg, eds., *World Mineral Exploration: Trends and Economic Issues* (Washington, D.C., Resources for the Future, forthcoming 1987).

[a]GDV = gross discovery value.

[b]Ratio = GDV/expenditure.

To sum up, the three studies highlighted in this section present success-ratio calculations based on broadly similar methodologies. Nevertheless, important differences exist in both methodology and data, some of which are brought out in a comparison of results (table 4–2). The Cranstone and Rose and Eggert studies, which use similar definitions of a discovery date, calculate several success ratios in recognition of the difficulties of quantifying gross discovery values. Cranstone's lower limit of success is based on discoveries that have been mined, are being mined, or are committed for production; the upper limit is based on all discoveries, plus multiplier additions to correct for systematic undervaluation of nonproducing—and mainly recent—discoveries. Rose and Eggert do not use multipliers to adjust discovery values but instead present three ratios: discoveries that have been mined, discoveries likely to produce in the next few years, and all discoveries identified by the study. Chazan calculates only point estimates. The French ratios are roughly comparable to the lower estimates of the other two studies because they are calculated from data on additions to reserves. Given the differences in methodology and the data uncertainties, the success ratios for the three studies are remarkably similar, ranging from ten to ninety-nine. Expressed in converse terms, exploration expenditures represent from 1 to 10 percent of the gross value of discovered metal, depending on how discovery value is defined.

While yielding information about discovery costs in relation to gross discovery values, success ratios fail to account for the net value of discoveries—that is, they do not consider metal recovery rates during milling and processing and capital and operating costs—and the time value of money. The study described next attempts to remedy these problems.

Discounted Net Returns from Exploration

Mackenzie and Woodall (forthcoming 1987) estimate and compare exploration productivity for base metals in Canada from 1946 to 1977 and Australia from 1955 to 1978. Productivity is calculated as the discounted net value of discoveries by comparing exploration expenditures with the net returns from development, mining, and processing of discoveries (assuming 1980 costs and mineral prices). Compared with the simpler productivity measures, the data requirements are much larger. The calculations are made on a before-tax basis.[1]

[1]Schreiber and Emerson (1984) have made similar, but not identical, discounted cash flow calculations that compare gold exploration productivity in Canada and the United States on an after-tax basis.

Table 4–2. International Comparisons of Success Ratios

Country	Success ratio[a] (gross discovery value/ exploration expenditure)
Canada, 1972–82[b]	
Deposits that have been mined, or are committed for production	22
All discoveries, plus multiplier additions	99
United States, 1970–83[c]	
Deposits that have been mined	10
Deposits likely to produce in the next few years	24
All discoveries	47
French exploration, 1973–82[d]	
French territory	16
Franc zone	52
Rest of world	20
Total	22

[a]Calculated in constant dollars or francs.
[b]Calculated from figure 4–3 and data in Donald A. Cranstone, "The Canadian Mineral Discovery Experience Since World War II," in John E. Tilton, Roderick G. Eggert, and Hans H. Landsberg, eds., *World Mineral Exploration: Trends and Economic Issues* (Washington, D.C., Resources for the Future, forthcoming 1987).
[c]From Arthur W. Rose and Roderick G. Eggert, "Exploration in the United States," in John E. Tilton, Roderick G. Eggert, and Hans H. Landsberg, eds., *World Mineral Exploration: Trends and Economic Issues* (Washington, D.C., Resources for the Future, forthcoming 1987).
[d]From table 4–1.

The Mackenzie and Woodall procedure is essentially a four-step process. The first step yields a time distribution of development and production cash flows for the average economic deposit in both Australia and Canada. The authors begin by choosing the subset of all mineral deposits discovered during the period that would be profitable under current economic conditions. To do this cash-flow distributions are calculated for the development and production periods of individual deposits, based on general estimates of mineral prices and smelter payment arrangements and deposit estimates for recoverable ore reserves, mill recovery rates, mine and mill capacities, capital costs, preproduction development periods, and operating costs. An economic deposit is defined as a discovery that has total revenues of at least $20 million (1980 Australian dollars) and at least a 10 percent rate of return. The time distribution of cash flows of the average economic deposit—the goal of the first step—is the average of all the economic deposits for the development and production stages of mineral supply. For Australia, 12 of 39 base metal discoveries satisfy the economic hurdle criteria, whereas 100 of 210 discoveries satisfy the criteria for Canada.[2]

[2]The numbers 39 and 210 represent what Mackenzie and Woodall classify as possible economic discoveries in Australia and Canada, respectively, and not the total number

The second step yields a time distribution of exploration expenditures necessary to discover the average economic deposit. The average exploration expenditure per economic discovery equals total exploration expenditures for the period in Australia or Canada divided by the total number of economic discoveries. The time distribution is obtained by, first of all, assuming an efficient annual exploration expenditure of A$2.5 million, and then dividing the efficient expenditure into the total exploration expenditure to arrive at the number of years of required exploration.[3]

The third step combines the cash-flow time distributions calculated in the first two steps into a time distribution of exploration, development, and production cash flows for the average economic deposit of Australia and Canada (figure 4–5). These cash-flow data then are used in base-case calculations of the productivity of base metals exploration (table 4–3). The results imply that base metals exploration in Australia has been less productive than similar exploration in Canada, as might have been expected before performing the calculations. Total exploration expenditures (in constant dollars) in Canada were twice those in Australia—roughly A$2 billion vs. A$1 billion—yet Canada has eight times as many economic discoveries. Canada's rate of return is more than twice Australia's rate, which actually reflects a negative discounted net return because the 8 percent rate of return is less than the discount rate of 10 percent. As for the reasons why productivity differs, Mackenzie and Woodall infer that either exploration is inherently more difficult in Australia than in Canada (due to deeply weathered surface geology in much of Australia) or that fewer base metal deposits occur in Australia (assuming similar levels of exploration technology and expertise in both countries). Furthermore, Canadian exploration benefits from lower capital and operating costs; several discoveries in Australia that were classified as uneconomic for purposes of the study would have been economic discoveries if located in an average area of Canada.

Although Australia's average economic discovery costs more than four times as much (undiscounted) and takes four times as long to discover as Canada's, Australia's deposit is some three times larger

---

of base metal deposits discovered in these two countries. Therefore, the Canadian number of 210 is not comparable to Cranstone's 900 Canadian discoveries, a more comprehensive number including many small deposits.

[3]This assumption of $2.5 million (1980 Australian dollars) as the most efficient annual exploration expenditure is a subjective judgment by Mackenzie and Woodall, based on discussions with explorationists and on observation of both successful and unsuccessful exploration organizations. A group of five to eight geologists with an annual budget of this size, it is argued, is small enough to retain many of the beneficial characteristics of entrepreneurial organizations and yet large enough to undertake a modern, scientific search.

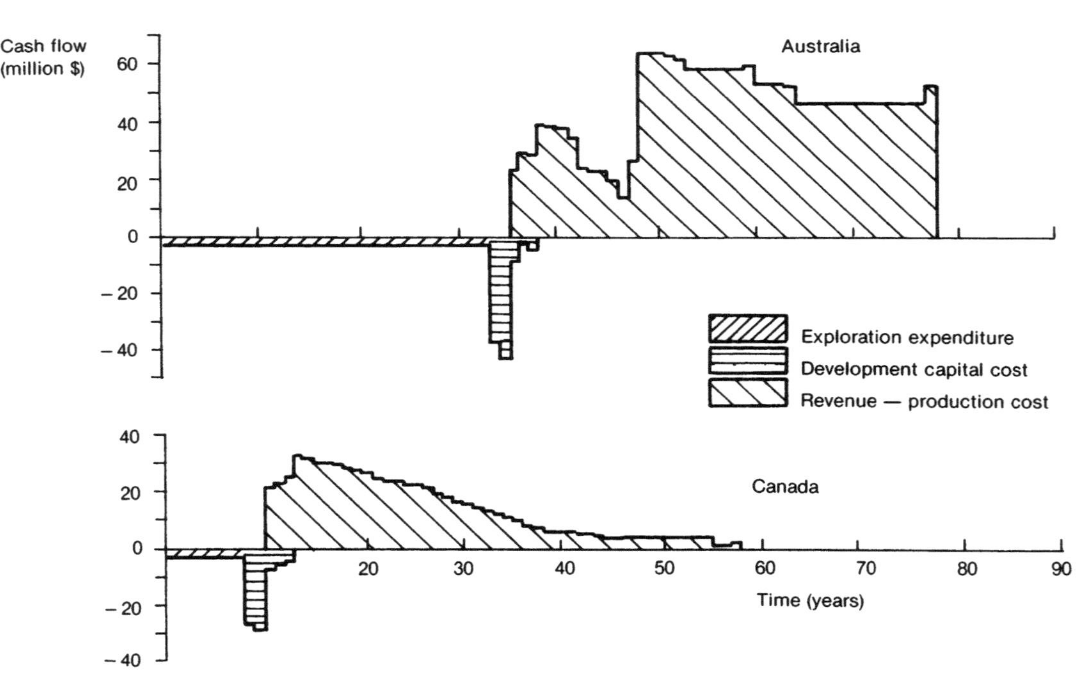

Figure 4–5. Time distribution of average cash flows for an economic deposit in Australia and Canada. *Source:* Brian Mackenzie and Roy Woodall, "Economic Productivity of Base Metal Exploration in Australia and Canada," in John E. Tilton, Roderick G. Eggert, and Hans H. Landsberg, eds., *World Mineral Exploration: Trends and Economic Issues* (Washington, D.C., Resources for the Future, forthcoming 1987) figure 11-7.

Table 4–3. **Base Metal Exploration Productivity: Base Case Conditions and Calculations for Australia and Canada**

	Australia	Canada
Time periods (years):		
Exploration	33	8
Development	5	5
Production	43	48
Undiscounted values (1980 A$ million)		
Revenue	3,431	1,137
Exploration expenditure	83	20
Development capital costs	97	71
Production costs	1,471	460
Discounted values (1980 A$ million; 10 percent real = discount rate)		
Average exploration expenditure	24	13
Average development and production return	9	64
Net return[a]	−15	51
Rate of return (percent)	8	20

Source: Based on data in Brian Mackenzie and Roy Woodall, "Economic Productivity of Base Metal Exploration in Australia and Canada," in John E. Tilton, Roderick G. Eggert, and Hans H. Landsberg, eds., *World Mineral Exploration: Trends and Economic Issues* (Washington, D.C., Resources for the Future, forthcoming 1987).

[a]Net return = (average development and production return) − (average exploration expenditure).

in terms of undiscounted revenue over the life of the project. The long average period of Australian exploration, however, yields much smaller discounted revenues. The long exploration period in Australia results directly from the assumption of an efficient annual exploration expenditure. If a larger efficient expenditure is assumed, the average period of exploration is shortened, production revenues move closer to the present, and thus discounted net returns increase not only for Australia but for Canada as well. If a smaller efficient expenditure is chosen, discounted net returns decline. Therefore, although the choice of an efficient annual exploration expenditure significantly influences the absolute level of exploration productivity for both countries, the differences in productivity between countries are much less dependent on the choice of an efficient annual expenditure.

The fourth step in the Mackenzie and Woodall methodology actually is a series of more detailed calculations that examine productivity in various geologic environments and over time and that test the sensitivity of the base-case calculations to metal price variations and different assumptions about an efficient annual exploration expenditure (mentioned earlier). Regarding different geologic environments, there is relatively little variability in productivity among geologic regions within the two countries compared to the intercountry differences. As for time trends, the data yield no clear trends in productivity, although productivity has declined slightly in both countries over time. Metal prices are an important determinant of exploration productivity; at higher prices, the number of economic deposits, discounted net values, and rates of return all increase; while at lower prices they all decline.

To sum up, measuring exploration productivity as the discounted net value of exploration activities remedies two serious drawbacks of the simpler productivity measures by (1) calculating output in net rather than gross terms, and (2) accounting for the time value of money. In defense of the simpler productivity calculations, the data requirements are much more onerous for the discounted value calculations. If the data were readily available, fewer researchers would have been satisfied with the simpler productivity ratios.

Nevertheless, even the discounted net value approach is not entirely satisfactory. First of all how much of what is calculated to be a return on an exploration investment might actually be an investment return on another factor of production such as development, mining, or mineral processing? And vice-versa? Without accurate measures of rates of return on investments in these postexploration activities, the exploration productivity number may be too high or too low. Suppose, for example, that a new mineral-processing technique reduces smelting costs over time or results in cost differences across countries. The discounted net value of exploration activities—that is, exploration productivity—will be higher in the time period or country with the new technique. This is not so much a difference in exploration productivity, however, as it is a difference in processing productivity.

This problem of distinguishing between exploration productivity and productivity in other stages of mineral supply is symptomatic of the more general problem of attempting to make broad inferences about mineral scarcity from information about mineral exploration. An often unstated assumption is that increasing discovery costs (implied by lower success ratios or lower discounted rates of return on exploration) will lead to increased scarcity—higher mineral prices, physical shortages, or both—ten to twenty years in the future. This is because exploration, it is alleged, is ultimately the only source of new minerals to replenish the dwindling stock of known deposits. Some economists go further and argue explicitly that increasing (decreasing) discovery costs indicate increasing (decreasing) scarcity under certain restrictive conditions. If depletion occurs, the argument goes, the economic rent[4] associated with known deposits should in-

[4]Economic rent is the revenue earned from a deposit in excess of what is required to bring it into production, or—if a deposit is in production already—in excess of what is required to keep it in production. See Tilton (1977) or a textbook in microeconomic theory such as Nicholson (1978) or Mansfield (1979).

crease (reflecting increased scarcity). Since it is the hope of capturing this rent that provides incentives for exploration and discovery in a competitive economy, exploration then should increase up to the point that the cost of discovering the next unit of metal (marginal cost, in economic terms) just equals the associated rent. Thus if scarcity and rents rise, marginal discovery costs also should rise.

However, using discovery cost as an indicator of economic rent and metal scarcity is fraught with both practical and theoretical difficulties. On a practical level, the available data on effectiveness of metals exploration do not test this hypothesis because they are not marginal discovery costs but rather are either ratios of metal value to exploration expenditure averaged over all deposits or rates of return on an average economic deposit. Theoretically, and more importantly, even if empirical estimates of marginal discovery costs were available, they very well might deviate from economic rent. The model hypothesizing a correlation between discovery costs and scarcity assumes that explorationists know the relationship between exploration inputs and outputs with certainty, whereas, in reality, this relationship is uncertain. Thus exploration levels and costs may be higher or lower than under conditions of certainty. Moreover, the model defines scarcity such that it cannot be mitigated by any cost-reducing effects of improved exploration and production technologies.[5]

The economic and other less formal attempts to infer something about mineral scarcity from information about exploration all ignore the important roles that other factors play in countering the cost-increasing effects of the depletion of known deposits. Advances in the processes and techniques of mining and mineral processing reduce costs and make it possible and profitable to mine and process mineral-bearing material in known deposits that previously had no economic value; such advances also influence the mineral targets sought by explorationists and, in turn, affect future discoveries. Material substitution replaces relatively scarce materials with relatively abundant ones. Finally, conservation and recycling, as well as shifts in consumption patterns, allow society to stretch out the use of known mineral deposits.

Therefore, calculations of exploration productivity, however sophisticated, provide useful information about the exploration phase of mineral supply—such as trends over time in the relationship between exploration expenditures and discovery values and intercountry or intermaterial comparisons of exploration effectiveness—but tell us remarkably little about overall mineral supply and resource scarcity.

[5]See Devarajan and Fisher (1982) and references cited therein for more detail about the theoretical relationship between mineral scarcity and discovery costs.

five

CONCLUSIONS

This study has been an economic and historical analysis of metallic mineral exploration over the last several decades. Chapter 1 set the stage with a brief description of the nature of mineral exploration and its role in mineral supply. Chapter 2 then used the limited amount of available quantitative data to identify and assess trends in exploration in selected countries. The chapter found that mineral prices importantly influence mineral exploration expenditures and explain many of the similarities among trends in countries outside the centrally planned economies. Mineral exploration expenditures and mineral prices tend to rise and fall together (although expenditures may lag prices by a year or two) because mineral price changes (1) reflect changes in the underlying conditions of mineral supply and demand, and in turn shape expectations about future prices and profits from exploration and mining, and (2) affect mining revenues and thus the availability and cost of capital for financing exploration. Differences among exploration trends in specific countries are found to be largely the result of differences in two country-specific factors: geologic potential and government policies (along with associated political risks).

Chapter 3 compared and contrasted exploration trends for iron ore, bauxite, copper, molybdenum, uranium, and gold and suggested that exploration and discovery of particular minerals or deposit types are generally episodic over time. Idealized episodes begin with incentives for stepped-up exploration for a particular mineral or deposit type; incentives, which make it potentially more profitable than before to discover an ore deposit, may include rising mineral prices and demand, new or revised geologic models that alter the way explorationists search for a mineral, or improved exploration and production technologies. Initial discoveries provide additional incentives for exploration over the short term by reducing geologic risk. Over the longer term, discoveries may discourage additional exploration by greatly expanding known sources of mineral supply; exploration also may decline as prices and demand fall. It goes without saying that there are many variations on this idealized scheme; the length, intensity, and regularity of cycles and the relative importance of various driving forces vary significantly from one mineral or deposit type to another.

Chapter 4 focused on the relationship between exploration expenditures and the results of mineral exploration. It found that exploration effectiveness is extremely difficult to quantify, both for empirical and conceptual reasons. Data on exploration expenditures, in many cases, are simply not available; when they are, they may not be consistent over time or across countries. Assigning values to mineral discoveries is, if anything, more difficult. Once exploration effectiveness has been quantified, it is difficult to know what can be properly inferred from this information.

Beyond the specific conclusions of the substantive chapters, the most important insights from this study all require treating exploration as an economic activity. As such it ultimately is motivated by the demand for minerals and their physical and chemical characteristics. Exploration is properly thought of as an investment, and organizations are motivated to invest in this activity by the desire for future economic gain however it may be defined; private corporations seek profits over the long term, some developing countries seek to promote overall economic development through mineral development, and centrally planned economies—such as the USSR—seek the lowest-cost mineral production for projected levels of consumption. Changes in the level and distribution of exploration, therefore, are determined by an interplay of technologic and political, as well as purely economic, factors that influence the expected benefits, costs, and risks associated with exploration.

Moreover, mineral exploration is only one of several ways that mining companies and society as a whole respond to the cost-increasing effects of the depletion of known deposits. These effects are brought on by moving to deposits that are, for example, lower grade, deeper below the surface, in more remote locations, or more complex metallurgically than deposits mined earlier. Other important responses include cost-saving advances in exploration and production techniques, material substitution, conservation, and recycling. Therefore any analysis of mineral exploration that fails to realize the important links among exploration activity and other aspects of mineral supply, as well as demand, is dangerously misleading.

STATISTICAL APPENDIX

Table A–1. Historical Mineral Exploration Expenditures, 1960–84
(millions of indicated national currencies)

Year	Australia		Canada		South Africa		United States		European companies	
	Current A$	1980/81 US$	Current C$	1980 US$	Current rands	1980 US$	Current US$	1980 US$	Current US$	1980 US$
1960	n.a.	n.a.	43.6	116.5	2.6	15.2	n.a.	n.a.	n.a.	n.a.
1961	n.a.	n.a.	43.5	115.9	3.7	21.4	60.9	167.3	n.a.	n.a.
1962	n.a.	n.a.	43.8	114.9	3.3	18.9	66.4	178.4	n.a.	n.a.
1963	n.a.	n.a.	43.5	112.1	2.5	13.9	71.6	189.4	n.a.	n.a.
1964	n.a.	n.a.	49.0	123.3	2.7	14.6	81.0	211.0	n.a.	n.a.
1965	14.0	56.1	54.0	131.6	3.5	18.4	86.3	218.8	n.a.	n.a.
1966	18.7	72.2	59.0	137.5	3.5	17.8	94.1	230.4	29.7	72.6
1967	25.5	96.1	52.8	118.5	3.7	18.2	103.5	247.1	41.4	98.9
1968	62.4	225.8	72.7	157.8	3.5	16.7	112.6	256.0	52.4	119.2
1969	97.0	336.8	96.6	201.5	4.3	19.4	129.3	278.4	77.7	167.4
1970	133.5	439.4	115.2	229.1	8.0	34.8	162.8	332.2	81.8	167.0
1971	95.9	293.6	86.4	166.4	12.5	51.4	166.6	321.6	99.6	192.3
1972	86.0	239.1	70.8	129.9	14.4	53.2	150.1	276.6	104.2	192.0
1973	87.3	211.5	84.0	141.1	19.3	59.9	192.6	333.5	123.0	213.0
1974	93.1	193.7	106.1	154.6	24.3	65.8	227.6	361.2	146.4	232.3
1975	84.9	154.2	119.7	157.5	29.1	71.8	220.0	317.9	184.2	266.3
1976	110.0	180.0	121.0	145.4	34.9	77.6	235.2	319.4	203.2/309.9	276.0/420.9
1977	135.6	204.5	164.2	183.6	39.8	79.4	271.2	345.3	382.5	487.1
1978	158.6	220.6	194.9	204.3	32.2	57.4	285.3	338.6	362.1	429.8
1979	255.6	322.8	248.9	236.5	34.9	54.2	327.2	356.8	420.0	458.0
1980	425.2	486.6	403.6	345.2	48.7	62.6	440.0	440.0	562.5	562.5
1981	531.0	551.4	484.7	374.8	71.2	83.7	484.1	441.4	515.5	470.0
1982	388.3	363.7	362.4	254.1	101.6	105.2	386.1	330.9	392.1	336.0
1983	388.0	338.2	300.0	199.4	102.8	93.1	314.7	259.8	358.0	295.6
1984	392.7	321.0	—	—	—	—	—	—	281.0	222.8

Note: n.a. = not available.

Sources and notes: **Australia**: Australian Bureau of Statistics, *Mineral Exploration Australia* (1984–86), Catalogue no. 8407.0; P. C. F. Crowson, "A Perspective on Worldwide Exploration for Minerals," in John E. Tilton, Roderick G. Eggert, and Hans H. Landsberg, eds., *World Mineral Exploration: Trends and Economic Issues* (Washington, D.C., Resources for the Future, forthcoming 1987). Data for 1965–67 are for calendar years, whereas data for 1968–84 represent fiscal years 1968/69–1984/85. Deflated using the Australian GDP implicit price index, 1980/81 = 100, then converted to U.S. dollars with the average exchange rate for 1980–81 of 1.1444 U.S. dollars per Australian dollar. Includes expensed and capitalized expenditures. Represents private expenditures outside of production leases. Excludes all government expenditures. Includes exploration for copper, lead, zinc, silver, nickel, cobalt, gold, iron ore, mineral sands, tin, tungsten, scheelite, wolfram, uranium, other metals, coal, construction materials, diamonds, and other materials.

Canada: Donald A. Cranstone, "The Canadian Mineral Discovery Experience Since World War II," in John E. Tilton, Roderick G. Eggert, and Hans H. Landsberg, eds., *World Mineral Exploration: Trends and Economic Issues* (Washington, D.C., Resources for the Future, forthcoming 1987); Hans W. Schreiber and Mark E. Emerson, "North American Hardrock Gold Deposits: An Analysis of Discovery Costs and the Cash Flow Potential," *Engineering and Mining Journal* vol. 185, no. 10, pp. 50–57, for 1983. Deflated using the Canadian GNP implicit price level, 1980 = 100, then converted to U.S. dollars with the average 1980 exchange rate of 1.1692 Canadian dollars per U.S. dollar. Includes off-site or general exploration. Excludes expenditures on properties in production or being prepared for production.

South Africa: Data collected by T. E. Beukes, "Mineral Exploration in South Africa," in John E. Tilton, Roderick G. Eggert, and Hans H. Landsberg, eds., *World Mineral Exploration: Trends and Economic Issues* (Washington, D.C., Resources for the Future, forthcoming 1987). Deflated using the South African GDP implicit price index, 1980 = 100, then converted to U.S. dollars with the average 1980 exchange rate of 1.2854 U.S. dollars per rand. Includes exploration for gold, platinum-group metals, copper, lead, zinc, tin, nickel, molybdenum, chromium, manganese, fluorspar, uranium, and coal.

United States: Based on a survey of thirty companies, Hans W. Schreiber and Mark E. Emerson, "North American Hardrock Gold Deposits," pp. 50–57. Deflated using the U.S. GNP implicit price index, 1980 = 100. Survey data were to include only expenditures for metals (excluding uranium) and industrial minerals, but some companies included an unknown amount for uranium and coal.

European companies: P. C. F. Crowson, "A Perspective on Worldwide Exploration for Minerals"; and personal communications for 1983 and 1984 data. All data derived from an annual survey of companies conducted by the Group of European Mining Companies from 1966 to 1978, and since then by the Comitée de Liaison des Industries de Métaux Non Ferreux. Data collected initially in terms of U.S. dollars, and then converted in this study to constant U.S. dollars using the U.S. GNP implicit price index, 1980 = 100. Excluded are expenditures for oil and gas exploration, but included are data for metals (including uranium), nonmetals, and coal. From 1966 to 1976, the following companies were included in the survey: BP Minerals, Charter Consolidated, Consolidated Gold Fields, Rio Tinto-Zinc, Selection Trust, Metallgesellschaft, Preussag, Uranerzbergbau, Urangesellschaft, Union Minière, Billiton, BRGM, Imetal, Société Nationale Elf Aquitane, Pechiney, Pertusola, and SAMI. Since then the survey also has included: Gewerkschaft Brunhilde, Saarberg-Interplan Uran GMBH, Greenex, Imperial Chemical Industries, Greek Mining Federation, and those members of the French Mining Federation not included previously. Figures are given in the table for both company bases for 1976.

Table A–2. Average Annual Mineral Prices, 1960–85

	Index of primary nonferrous metals		Copper		Lead		Zinc		Gold		Silver		Uranium	
Year	Current US$	1980 US$ (1980 = 100)	Current US$	1980 US$	Current US$	1980 US$	Current US$	1980 US$	Current US$	1980 US$	Current US$	1980 US$	Current US$	1980 US$
1960	83.8	61.9	32.1	88.9	12.0	33.1	13.0	35.9	35.00	97.07	0.91	2.52	8.75	24.27
1961	81.7	59.7	29.9	82.2	10.9	29.9	11.5	31.7	35.00	96.14	0.92	2.53	8.50	23.35
1962	81.4	58.2	30.6	82.2	9.6	25.9	11.6	31.2	35.00	94.03	1.09	2.93	8.15	21.90
1963	82.6	58.2	30.6	80.9	11.1	29.5	12.0	31.7	35.00	92.58	1.28	3.39	7.82	20.68
1964	90.0	62.4	32.0	83.3	13.6	35.4	13.6	35.3	35.00	91.17	1.29	3.36	8.00	20.84
1965	96.8	65.3	35.0	88.8	16.0	40.6	14.5	36.8	35.00	88.74	1.29	3.27	8.00	20.28
1966	96.0	62.6	36.2	88.6	15.1	37.0	14.5	35.5	35.00	85.70	1.29	3.16	8.00	19.59
1967	100.0	63.6	38.2	91.3	14.0	33.4	13.8	33.0	35.00	83.55	1.55	3.70	8.00	19.10
1968	106.7	64.6	41.9	95.1	13.2	30.0	13.5	30.7	39.26	89.25	2.14	4.86	6.50	14.78
1969	114.2	65.5	47.5	102.3	14.9	32.1	14.6	31.4	41.51	89.38	1.79	3.85	6.23	13.41
1970	128.1	69.6	57.7	117.7	15.6	31.9	15.3	31.3	36.41	74.29	1.77	3.61	6.23	12.71
1971	117.5	60.4	51.4	99.3	13.8	26.6	16.1	31.1	41.25	79.62	1.55	2.99	6.00	11.58
1972	115.6	56.7	50.6	93.3	15.0	27.7	17.8	32.7	58.60	108.00	1.68	3.10	5.95	10.97
1973	139.1	64.1	58.9	101.9	16.3	28.2	20.7	35.8	97.81	169.34	2.56	4.43	6.75	11.69
1974	197.7	83.5	76.7	121.6	22.5	35.8	36.0	57.1	159.74	253.51	4.71	7.47	12.75	20.23
1975	184.8	71.1	63.5	91.8	21.5	31.1	39.0	56.3	161.49	233.38	4.42	6.39	28.50	41.19
1976	191.6	69.3	68.8	93.5	23.1	31.4	37.0	50.3	125.32	170.20	4.35	5.91	40.50	55.01
1977	206.0	69.8	65.8	83.8	30.7	39.1	34.4	43.8	148.31	188.86	4.62	5.88	42.75	54.44
1978	218.3	69.0	65.5	77.8	33.7	39.9	31.0	36.8	193.55	229.74	5.40	6.41	43.30	51.40
1979	295.1	85.7	92.3	100.7	52.6	57.4	37.3	40.7	307.50	335.28	11.09	12.09	41.88	45.66

1980	375.6	100.0	101.4	101.4	42.6	42.6	37.4	37.4	612.56	612.56	20.63	20.63	29.75	29.75
1981	326.1	79.2	83.7	76.3	35.5	32.4	44.6	40.6	459.64	419.05	10.52	9.59	23.90	21.79
1982	284.3	64.9	72.9	62.5	25.5	21.9	38.5	33.0	375.91	322.15	7.95	6.81	19.75	16.93
1983	303.4	66.7	77.9	64.3	21.7	17.9	41.4	34.2	424.00	350.07	11.44	9.45	24.00	19.82
1984	288.1	60.8	66.8	52.9	25.6	20.3	48.6	38.5	360.66	285.93	8.14	6.45	18.60	14.75
1985	266.1	54.4	65.6	50.3	19.1	14.6	40.4	31.0	318.00	243.98	6.14	4.71	16.00	12.28

Sources and notes: **Primary nonferrous metals**: Price index from U.S. Department of Labor, *Producer Prices and Price Indexes* (Washington, D.C., Government Printing Office, various years). Industry Code no. 1022. 1967 = 100 for current dollar index. 1980 = 100 for constant dollar index.

Copper: Cents/pound, U.S. domestic refinery price. "Average annual metal prices 1925–1985," *Engineering and Mining Journal* vol. 187, no. 3, p. 27.

Lead: Cents/pound, U.S. primary delivered price. "Average annual metal prices 1925–1985," p. 27.

Zinc: Cents/pound, U.S. high grade delivered price. "Average annual metal prices 1925–1985," p. 27.

Gold: Dollars/ounce, annual average selling price. 1960–79: U.S. Bureau of Mines, *Mineral Facts and Problems*, Bulletin 671 (Washington, D.C., Government Printing Office, 1980). 1980–85: U.S. Bureau of Mines, *Mineral Commodity Summaries 1986* (Washington, D.C., Government Printing Office, 1980).

Silver: Dollars/ounce. "Average annual metal prices 1925–1985," p. 27.

Uranium: Dollars/pound uranium oxide. 1960–83: Thomas Neff, *The International Uranium Market* (Cambridge, Mass., Ballinger Publishing Company, 1984). 1984–85: "Average annual metal prices 1925–1985," p. 27.

All prices deflated using the U.S. GNP implicit price index, 1980 = 100.

Table A–3. Uranium Exploration Expenditures by European Companies and in the United States, 1966–84
(million US$)

	European companies		United States			
			Dept. of Energy		OECD/IAEA	
Year	Current US$	1980 US$	Current US$	1980 US$	Current US$	1980 US$
1966	0.6	1.5	8.4	20.6	n.a.	n.a.
1967	2.1	5.0	24.8	59.2	n.a.	n.a.
1968	11.9	27.1	53.4	121.4	n.a.	n.a.
1969	16.6	35.7	58.7	126.4	n.a.	n.a.
1970	20.3	41.4	52.2	106.5	n.a.	n.a.
1971	15.9	30.7	41.2	79.5	n.a.	n.a.
1972	14.8	27.3	32.4	59.7	32.4	59.7
1973	28.6	49.5	49.5	85.7	49.6	85.9
1974	45.4	72.1	79.1	125.5	80.9	128.4
1975	61.1	88.3	122.0	176.3	130.2	188.2
1976	65.3/160.6	88.7/218.1	170.7	231.8	195.9	266.1
1977	208.6	265.6	258.1	328.7	293.5	373.7
1978	207.6	246.4	314.3	373.1	371.5	441.0
1979	224.2	244.5	315.9	344.4	385.6	420.4
1980	251.6	251.6	267.0	267.0	332.6	332.6
1981	176.7	161.1	144.8	132.0	180.3	164.4
1982	139.7	119.7	73.6	63.1	97.8	83.8
1983	111.1	91.7	38.9	32.1	56.7	46.8
1984	78.7	62.4	26.5	21.0	n.a.	n.a.

Note: In 1976 and subsequent years, the number of European companies in the survey has been larger than in the previous years. Expenditures are given in the table in both company bases for 1976. See notes to appendix table A–1. n.a. = not available.

Sources: European companies: 1966–82, P. C. F. Crowson, "A Perspective on Worldwide Exploration for Minerals," in John E. Tilton, Roderick G. Eggert, and Hans H. Landsberg, *World Mineral Exploration: Trends and Economic Issues* (Washington, D.C., Resources for the Future, forthcoming 1987); 1983–84, personal communication. United States: (a) Dept. of Energy, 1966–82, P. C. F. Crowson, "A Perspective on Worldwide Exploration for Minerals," in John E. Tilton, Roderick G. Eggert, and Hans H. Landsberg, *World Mineral Exploration: Trends and Economic Issues* (Washington, D.C., Resources for the Future, forthcoming 1987); 1983–84, Energy Information Administration, U.S. Department of Energy, *Uranium Industry Annual 1984*, DOE/EIA-0478(84), (Washington, D.C., Government Printing Office, 1985); (b) OECD Nuclear Energy Agency and the International Atomic Energy Agency, *Uranium Resources, Production, and Demand* (Paris, Organisation for Economic Co-operation and Development, 1977, 1979, 1982, 1983, 1984).

Table A–4. OECD-IAEA Estimates of Uranium Exploration Expenditures for Selected Countries and the World Outside the Centrally Planned Economies, 1972–83
(million US$)

	World[a]		Australia[b]		Brazil		Canada[c]		France		Gabon[d]		South Africa	
Year	Current US$	1980 US$	Current US$	1980 US$	Current US$	1980 US$	Current US$	1980 US$	Current US$	1980 US$	Current US$	1980 US$	Current US$	1980 US$
1972	72.8	134.2	12.8	23.6	1.5	2.8	6.0	11.1	4.8	8.8	4.0	7.4	—	—
1973	104.2	180.4	16.3	28.2	8.5	14.7	6.0	10.4	5.4	9.3	4.0	6.9	—	—
1974	145.7	231.2	14.6	23.2	10.0	15.9	8.0	12.7	8.2	13.0	4.0	6.3	0.8	1.3
1975	228.9	330.8	10.0	14.5	12.0	17.3	23.5	34.0	14.6	21.1	4.0	5.8	3.4	4.9
1976	346.0	469.9	14.0	19.0	24.4	33.1	43.1	58.5	19.6	26.6	4.0	5.4	17.3	23.5
1977	485.0	617.6	18.0	22.9	25.6	32.6	65.5	83.4	18.2	23.2	4.8	6.1	12.3	15.7
1978	618.8	734.5	29.0	34.4	29.3	34.8	75.9	90.1	26.7	31.7	8.5	10.1	24.6	29.2
1979	703.5	767.0	32.0	34.9	12.5	13.6	110.9	120.9	61.7	67.3	9.6	10.5	31.3	34.1
1980	682.5	682.5	45.0	45.0	8.6	8.6	109.5	109.5	89.5	89.5	8.6	8.6	22.1	22.1
1981	481.2	438.7	43.0	39.2	18.8	17.1	85.2	77.7	70.5	64.3	8.3	7.6	19.2	17.5
1982	286.6	245.6	31.0	26.6	17.9	15.3	57.5	49.3	55.7	47.7	7.7	6.6	6.8	5.8
1983[e]	189.9	156.8	15.0	12.4	4.3	3.6	34.5	28.5	50.7	41.9	5.9	4.9	5.3	4.4

Source: OECD Nuclear Energy Agency and the International Atomic Energy Agency, *Uranium Resources, Production, and Demand* (Paris, Organisation for Economic Co-operation and Development, 1977, 1979, 1982, 1983), except where noted otherwise.

[a]World outside the centrally planned economies.

[b]1982 and 1983 figures are interpolations from fiscal-year data in Australian Bureau of Statistics, *Mineral Exploration Australia (1984, 1985)*, Catalogue no. 8407.0.

[c]1971–74 total allocated across individual years.

[d]Pre-1977 total: allocated across individual years, and from P. C. F. Crowson, "A Perspective on Worldwide Exploration for Minerals," in John E. Tilton, Roderick G. Eggert, and Hans H. Landsberg, *World Mineral Exploration: Trends and Economic Issues* (Washington, D.C., Resources for the Future, forthcoming 1987).

[e]Planned.

Table A–5. Gold Exploration Expenditures in Selected Countries and by French Organizations, 1960–83
(millions of indicated national currency)

	United States			Canada			Australia			South Africa			French exploration	
Year	Current US$	1980 US$	Share[a]	Current US$	1980 US$	Share[a]	Current A$	1980–81 A$	Share[a]	Current rand	1980 rand	Share[a]	1982 FF	Share[a]
1960	n.a.	n.a.	n.a.	n.a.	n.a.	n.a.	n.a.	n.a.	n.a.	1.5	6.8	58	n.a.	n.a.
1961	0.6	1.6	1	4	11	9	n.a.	n.a.	n.a.	2.4	10.8	66	n.a.	n.a.
1962	0.7	1.9	1	4	11	9	n.a.	n.a.	n.a.	2.7	12.0	81	n.a.	n.a.
1963	1.4	3.7	2	4	11	9	n.a.	n.a.	n.a.	1.6	6.9	65	n.a.	n.a.
1964	1.6	4.2	2	5	13	10	n.a.	n.a.	n.a.	1.8	7.6	64	n.a.	n.a.
1965	2.6	6.6	3	5	13	9	n.a.	n.a.	n.a.	2.1	8.6	60	n.a.	n.a.
1966	2.8	6.9	3	5	12	9	n.a.	n.a.	n.a.	1.9	7.5	54	n.a.	n.a.
1967	4.1	9.8	4	7	17	10	n.a.	n.a.	n.a.	1.8	6.9	49	n.a.	n.a.
1968	5.6	12.7	5	10	23	15	n.a.	n.a.	n.a.	1.7	6.3	50	n.a.	n.a.
1969	7.8	16.8	6	18	39	20	n.a.	n.a.	n.a.	2.6	9.1	62	n.a.	n.a.
1970	9.8	20.0	6	18	37	19	n.a.	n.a.	n.a.	3.8	12.9	47	n.a.	n.a.
1971	11.7	22.6	7	14	27	19	n.a.	n.a.	n.a.	4.3	13.8	34	n.a.	n.a.
1972	12.0	22.1	8	17	31	20	n.a.	n.a.	n.a.	2.5	7.2	17	n.a.	n.a.
1973	19.3	33.4	10	17	29	20	n.a.	n.a.	n.a.	2.9	7.0	15	11	2
1974	27.3	43.3	12	22	35	20	n.a.	n.a.	n.a.	2.8	5.9	12	15	2
1975	30.8	44.5	14	32	46	25	n.a.	n.a.	n.a.	5.5	10.6	19	39	5
1976	40.0	54.3	17	33	45	25	n.a.	n.a.	n.a.	6.8	11.8	20	24	3
1977	54.2	69.0	20	42	53	25	n.a.	n.a.	n.a.	9.7	15.1	24	15	2
1978	65.6	77.9	23	58	69	30	n.a.	n.a.	n.a.	10.9	15.1	34	19	2
1979	91.6	99.9	28	87	95	35	29.93	33.12	15	12.3	14.9	35	25	3
1980	145.2	145.2	33	141	141	35	69.24	69.24	20	17.9	17.9	37	35	4
1981	193.6	176.5	40	170	155	35	94.83	86.05	24	26.4	24.2	37	58	5
1982	181.4	155.5	47	181	155	50	96.35	78.85	30	38.2	30.8	38	97	9
1983	173.1	142.9	55	195	161	65	149.34	113.74	44	44.9	31.6	44	n.a.	n.a.

Sources and notes: n.a. = not available.

United States: Hans W. Schreiber and Mark E. Emerson: "North American Hardrock Gold Deposits: An Analysis of Discovery Costs and the Cash Flow Potential," *Engineering and Mining Journal*, vol. 185, no. 10, pp. 50–57. Mineral exploration universe is supposed to include only nonfuel minerals and industrial minerals, but an unknown amount of uranium and coal exploration is included. Based on survey data from thirty companies. See notes to appendix table A–1.

Canada: Hans W. Schreiber and Mark E. Emerson, "North American Hardrock Gold Deposits." Mineral universe includes metals and industrial minerals (see note above).

South Africa: T. E. Beukes, "Mineral Exploration in South Africa," in John E. Tilton, Roderick G. Eggert, and Hans H. Landsberg, eds., *World Mineral Exploration: Trends and Economic Issues* (Washington, D.C., Resources for the Future, forthcoming 1987). Data include exploration for platinum-group metals as well as gold. The mineral universe includes coal, uranium, copper, lead, zinc, tin, nickel, molybdenum, chromium, manganese, and fluorspar. See notes to appendix table A–1.

Australia: Australian Bureau of Statistics, *Mineral Exploration Australia (1984–1986)*, Catalogue no. 8407.0. P. C. F. Crowson, "A Perspective on Worldwide Exploration for Minerals," in John E. Tilton, Roderick G. Eggert, and Hans H. Landsberg, *World Mineral Exploration: Trends and Economic Issues* (Washington, D.C., Resources for the Future, forthcoming 1987). Data are for fiscal years 1979/80–1983/84. Mineral universe includes metals, industrial minerals, coal, and diamonds. See notes to appendix table A–1.

French exploration: Willy Chazan, "French Mineral Exploration, 1973–1982," in John E. Tilton, Roderick G. Eggert, and Hans H. Landsberg, eds., *World Mineral Exploration: Trends and Economic Issues* (Washington, D.C., Resources for the Future, forthcoming 1987).

[a]Share of total exploration expenditures in a particular country, or of French exploration (percentage). See notes above for the relevant mineral universe.

REFERENCES

Those references marked with asterisks are for readers wishing more detail on the general nature of mineral exploration.

*Adelman, M. A. 1970. "Economics of Exploration for Petroleum and Other Minerals," *Geoexploration* vol. 8, pp. 131–150. An economist's view of exploration and how it fits into mineral supply.

Astakhov, Alexander S., Michail N. Denisov, and Vladimir K. Pavlov. forthcoming 1987. "Prospecting and Exploration in the Soviet Union," in John E. Tilton, Roderick G. Eggert, and Hans H. Landsberg, eds., *World Mineral Exploration: Trends and Economic Issues* (Washington, D.C., Resources for the Future).

Australian Bureau of Statistics. 1984–1986, annual. *Mineral Exploration Australia*, Catalogue no. 8407.0.

*Bailly, Paul A. 1972. "Mineral Exploration Philosophy," *American Mining Congress Journal* vol. 58, no. 4, pp. 31–37. A professional explorationist's view of corporate exploration philosophies.

———. 1979a. "Managing for Ore Discoveries," *Mining Engineering* vol. 31, no. 6, pp. 663–671.

*———. 1979b. "The Role of Luck in Mineral Exploration," *American Mining Congress Journal* vol. 65, no. 4, pp. 56–61.

Barber, G. A., and Siegfried Muessig. 1984. "Minerals Exploration Statistics for the Years 1980, 1981, and 1982," *Economic Geology* vol. 79, no. 7, pp. 1,768–1,776.

———, ———. 1985. "1983 Minerals Exploration Statistics: United States and Canadian Companies," *Economic Geology* vol. 80, no. 7, pp. 2,060–2,066.

———, ———. 1986. "1984 Minerals Exploration Statistics: United States and Canadian Companies," *Economic Geology*, in press.

Barnett, D. W. 1980. "Australia's Minerals Policy." Preprint of paper presented at the joint meeting of the Institution of Mining and Metallurgy, the Society of Mining Engineers of AIME, and the Metallurgical Society of AIME, London, England.

Behling, David J., Jr., Richard S. Dobias, and Norma J. Anderson. 1985. *1984 Capital Investments in the World Petroleum Industry* (New York, Chase Manhattan Bank).

Bennethum, G., and L. C. Lee. 1975. "Is Our Account Overdrawn?" *American Mining Congress Journal* vol. 61, no. 9, pp. 33–48.

Beukes, T. E. forthcoming 1987. "Mineral Exploration in South Africa," in John E. Tilton, Roderick G. Eggert, and Hans H. Landsberg, eds., *World Mineral Exploration: Trends and Economic Issues* (Washington, D.C., Resources for the Future).

Branson, William H. 1979. *Macroeconomic Theory and Policy*, 2nd ed. (New York, Harper and Row).

Brown, William K. 1983. "Exploration for Gold: Costs and Results." Preprint no. 83-142. Paper presented at the SME-AIME Annual Meeting, Atlanta, Georgia, March 6–10.

Chazan, Willy. forthcoming 1987. "French Mineral Exploration, 1973–82," in John E. Tilton, Roderick G. Eggert, and Hans H. Landsberg, eds., *World Mineral Exploration: Trends and Economic Issues* (Washington, D.C., Resources for the Future).

Cook, Douglas R. 1983. "Exploration or Acquisition: The Options for Acquiring Mineral Deposits," *American Mining Congress Journal* vol. 69, no. 21, pp. 10–12.

Cranstone, Donald A. forthcoming 1987. "The Canadian Mineral Discovery Experience since World War II," in John E. Tilton, Roderick G. Eggert, and Hans H. Landsberg, eds., *World Mineral Exploration: Trends and Economic Issues* (Washington, D.C., Resources for the Future).

Crowson, P. C. F. forthcoming 1987. "A Perspective on Worldwide Exploration for Minerals," in John E. Tilton, Roderick G. Eggert, and Hans H. Landsberg, eds., *World Mineral Exploration: Trends and Economic Issues* (Washington, D.C., Resources for the Future).

Derry, Duncan R., and James K. B. Booth. 1978. "Mineral Discoveries and Exploration Expenditure—A Revised Review 1966–1976," *Mining Magazine* vol. 138, no. 5, pp. 430–433.

Devarajan, Shantayanan, and Anthony C. Fisher. 1982. "Measures of Natural Resource Scarcity under Uncertainty," in V. Kerry Smith and John V. Krutilla, eds., *Explorations in Natural Resource Economics* (Baltimore, Md., Johns Hopkins University Press for Resources for the Future).

DeYoung, John H., Jr. 1977. "Effect of Tax Laws on Mineral Exploration in Canada," *Resources Policy* vol. 3, no. 2, pp. 96–107.

______. 1978. "Measuring the Economic Effects of Tax Laws on Mineral Exploration," *Proceedings of the Council of Economics of AIME*, 107th annual meeting, pp. 29–40.

Eggert, Roderick G. 1985a. "Exploration's Role in Iron and Aluminum Supply Since the Second World War," *Natural Resources Forum* vol. 9, no. 3, pp. 187–195.

______. 1985b. "Mineral Exploration in the USSR and the USA," *Resources Policy* vol. 11, no. 2, pp. 128–140.

———. forthcoming 1987. "Base and Precious Metals Exploration by Major Corporations," in John E. Tilton, Roderick G. Eggert, Hans H. Landsberg, eds., *World Mineral Exploration: Trends and Economic Issues* (Washington, D.C., Resources for the Future).

Energy Information Administration, U.S. Department of Energy. 1985. *Uranium Industry Annual 1984*. DOE/EIA-0478(84). (Washington, D.C., Government Printing Office).

Engineering and Mining Journal. 1986. "Average Annual Metal Prices 1925–1985" vol. 187, no. 3, p. 27.

Gilmour, P. 1982. "Grades and Tonnages of Porphyry Copper Deposits," in S. R. Titley, ed., *Advances in Geology of the Porphyry Copper Deposits* (Tucson, Ariz., University of Arizona Press).

Harris, D. P., and B. J. Skinner. 1982. "The Assessment of Long-Term Supplies of Minerals," in V. K. Smith and J. V. Krutilla, eds., *Explorations in Natural Resource Economics* (Baltimore Md., Johns Hopkins University Press for Resources for the Future).

Hodge, B. L., and O. L. Oldham. 1979. "Mineral Exploration in 1978," *World Mining* vol. 32, no. 8, pp. 89–99.

Johnson, Charles J., and Allen L. Clark. forthcoming 1987. "Mineral Exploration in Developing Countries: Botswana and Papua New Guinea Case Studies," in John E. Tilton, Roderick G. Eggert, and Hans H. Landsberg, eds., *World Mineral Exploration: Trends and Economic Issues* (Washington, D.C., Resources for the Future).

King, Haddon F. 1973. "A Look at Mineral Exploration 1934–1973," *Proceedings of the Australasian Institute of Mining and Metallurgy, Winter Annual Conference*, pp. 1–16.

Kramer, D. 1983. "Gold Fields is Focusing Exploration in N. America," *American Metal Market*, September 14, pp. 1, 20.

Mackenzie, Brian, and Roy Woodall. forthcoming 1987. "Economic Productivity of Base Metal Exploration in Australia and Canada," in John E. Tilton, Roderick G. Eggert, and Hans H. Landsberg, eds., *World Mineral Exploration: Trends and Economic Issues* (Washington, D.C., Resources for the Future).

Manners, Gerald. 1971. *The Changing World Market for Iron Ore 1950–1980* (Baltimore, Md., Johns Hopkins University Press for Resources for the Future).

Mansfield, Edwin. 1979. *Microeconomics: Theory and Applications*, 3rd ed. (New York, W. W. Norton).

*Miller, L. J. 1976. "Corporations, Ore Discovery, and the Geologist," *Economic Geology* vol. 71, pp. 836–847. Describes the discovery of the Kidd Creek (Ontario, Canada) copper-zinc-silver deposit.

Moran, T. H. 1973. "Transnational Strategies of Protection and Defense by Multinational Corporations: Spreading the Risk and Raising the Cost for Nationalization in Natural Resources," *International Organization* vol. 27, no. 2, pp. 273–287.

Mullins, W. J., R. D. Lawrence, and D. W. Deschamps. 1977. "Explorationists Seek U_3O_8, Porphyry, Silver, Sulphide Deposits," *World Mining* vol. 30, no. 7, pp. 104–111, 235.

Neff, Thomas. 1984. *The International Uranium Market* (Cambridge, Mass., Ballinger Publishing Company).

*Newendorp, Paul D. 1975. *Decision Analysis for Petroleum Exploration* (Tulsa, Oklahoma, The Petroleum Publishing Company). Applies statistical decision theory to risk and uncertainty in exploration decisionmaking.

Nicholson, Walter. 1978. *Microeconomic Theory: Basic Principles and Extensions*, 2nd ed. (Hinsdale, Ill., The Dryden Press).

OECD Nuclear Energy Agency and the International Atomic Energy Agency. 1977, 1979, 1982, 1983, 1984. *Uranium Resources, Production and Demand* (Paris, Organisation for Economic Co-operation and Development).

*Office of Technology Assessment, Congress of the United States. 1979. *Management of Fuel and Nonfuel Minerals in Federal Land: Current Issues and Status* (Washington, D.C., Government Printing Office). Chapter 2 provides a comprehensive technical description of exploration intended for the nonspecialist.

Prain, Ronald. 1975. *Copper: The Anatomy of an Industry* (London, Mining Journal Books Limited).

*Preston, Lee E. 1960. *Exploration for Nonferrous Metals* (Washington, D.C., Resources for the Future). Another economist's viewpoint, with good discussion of pre–World War II exploration.

Radetzki, Marian. 1982. "Has Political Risk Scared Mineral Investment away from the Deposits in Developing Countries?" *World Development* vol. 10, no. 1, pp. 39–48.

Rose, Arthur W., and Roderick G. Eggert. forthcoming 1987. "Exploration in the United States," in John E. Tilton, Roderick G. Eggert, and Hans H. Landsberg, eds., *World Mineral Exploration: Trends and Economic Issues* (Washington, D.C., Resources for the Future).

Schreiber, Hans W., and Mark E. Emerson. 1984. "North American Hardrock Gold Deposits: An Analysis of Discovery Costs and the Cash Flow Potential," *Engineering and Mining Journal* vol. 185, no. 10, pp. 50–57.

*Snow, Geoffrey C., and Brian W. Mackenzie. 1981. "The Environment of Exploration: Economic, Organizational, and Social Constraints," *Economic Geology*, 75th Anniversary Volume, pp. 871–896.

Sullivan, C. J. 1974. "Mineral Investment Decisions for the Coming Decades," *CIM Bulletin* vol. 67, no. 747, pp. 19–22.

Tilton, John E. 1977. *The Future of Nonfuel Minerals* (Washington, D.C., The Brookings Institution).

______. Roderick G. Eggert, and Hans H. Landsberg, editors. forthcoming 1987. *World Mineral Exploration: Trends and Economic Issues* (Washington, D.C., Resources for the Future).

U.S. Bureau of Mines. 1974–1983. *Minerals Yearbook*, vol. 1, (Washington, D.C., Government Printing Office).

———. 1980. *Mineral Facts and Problems*, 1980 ed. Bulletin 671. (Washington, D.C., Government Printing Office).

———. 1986. *Mineral Commodity Summaries 1986* (Washington, D.C., Government Printing Office).

U.S. Department of Labor. *Producer Prices and Price Indexes* (Washington, D.C., Government Printing Office, various years).

Walthier, T. N. 1976. "The Shrinking World of Exploration, Part 2," *Mining Engineering* vol. 28, no. 5, pp. 46–50.

www.ingramcontent.com/pod-product-compliance
Ingram Content Group UK Ltd.
Pitfield, Milton Keynes, MK11 3LW, UK
UKHW020425010325
455677UK00029B/1007